Photovoltaism, Agriculture and Ecology

Series Editor
Françoise Gaill

Photovoltaism, Agriculture and Ecology

From Agrivoltaism to Ecovoltaism

Claude Grison
Lucie Cases
Mailys Le Moigne
Martine Hossaert-McKey

iSTE

WILEY

First published 2021 in Great Britain and the United States by ISTE Ltd and John Wiley & Sons, Inc.

ISTE Ltd
27-37 St George's Road
London SW19 4EU
UK

www.iste.co.uk

John Wiley & Sons, Inc.
111 River Street
Hoboken, NJ 07030
USA

www.wiley.com

Library of Congress Control Number: 2021942790

British Library Cataloguing-in-Publication Data
A CIP record for this book is available from the British Library
ISBN 978-1-78630-720-0

MIX
Paper from responsible sources
FSC® C013604

Contents

Foreword by Yvon Le Maho

This update by Claude Grison, Lucie Cases, Mailys Le Moigne and Martine Hossaert-McKey on how the development of a renewable energy, photovoltaics, could be part of a real ecological transition meets a real need.

Indeed, in view of the two major environmental issues of our time – the fight against climate change and the reduction of biodiversity – we are faced with what appears to be an abuse of language that hides a real divide. What is it about? In the dominant discourse, climate change is associated with an "ecological catastrophe". Certainly, according to the most probable scenarios, the announced global warming will eventually result in a significant loss of biodiversity. To avoid such an outcome, this implies that we must fight against climate change.

In fact, the seriousness of this prognosis is largely linked to the fact that climate change will weigh on an already weakened biodiversity. For most of the reduction in biodiversity already observed, more than 80% of the decrease in populations of threatened species, according to a recent WWF study, has nothing to do with climate change. Land artificialization, overexploitation of resources, intensive farming and disturbance are the main causes. It is therefore two parallel actions, both fighting against the human factors of climate change, that is, against the increase of greenhouse gases, and recovering biodiversity, that must be carried out. The indicators of the better health of fauna that followed the recent period of confinement caused by the COVID-19 health crisis have shown something important: the reduction of the disturbance induced by humans towards the animal populations during their period of reproduction has already shown very positive effects… However, the stakes of the fight against the reduction of biodiversity are not only patrimonial but also economic. The crucial importance of pollinating insects is enough to remind us of this.

In addition, there is a divide between scientific and technological research that is moving forward and an ecology that can easily be derided as a movement promoting a return to the past. It is only a softened form of the parable in which he who wants to get rid of his dog cries out that he has rabies… It is true that the indistinguishability in terms between a militant and political ecology and scientific ecology does not make things easier. But this dichotomy between research that is synonymous with progress and an ecology that would be a brake on it can be found right up to the ministerial level in the main European countries. It is even the main obstacle to the development of expertise on biodiversity, as we noted with Julien Boucher, *Maître des requêtes*[1] at the *Conseil d'État* (Council of State), in a report we wrote commissioned by the then Minister of the Environment, Jean-Louis Borloo, and the Secretary of State, Chantal Jouanno, a report that we submitted to Nathalie Kosciusko-Morizet, who succeeded Borloo as Minister of Ecology.

It is therefore in more than one way that the book by Claude Grison and her colleagues meets a real need. As these authors clearly show, scientific and technological developments and advances in engineering must not be carried out separately to meet the challenges of climate and biodiversity. Optimization must be sought through an approach that can only be multidisciplinary and interdisciplinary in order to reconcile the development of photovoltaics with the need to save water reserves and halt the decline in animal and plant biodiversity. In order to promote this multidisciplinary and interdisciplinary approach, the authors of this book obviously had many assets at their disposal. They are breaking down the barriers of science at the interface between chemistry and ecology, an approach that they consider indispensable to the ecological transition. Claude Grison serves as a Director of Research at the CNRS (French National Center for Scientific Research) and leads the Bio-inspired Chemistry and Ecological Innovations Laboratory, a laboratory that brings together chemists and ecologists around common themes. She has associated two students from the laboratory, Lucie Cases and Mailys Le Moigne, who share her passion for the search for ecological and innovative solutions in the production of new energy. Through her expertise in chemical ecology, Martine Hossaert-McKey,

1 Master of Requests, i.e. a high judicial officer of administrative law in France.

Emeritus Director of Research at the CNRS in charge of biodiversity, makes the link between the plant and insect worlds by studying how chemical language shapes these interactions and structures ecosystems. The authors of this book have taken great care to show not only the advantages but also the limitations of the different techniques implemented in agrivoltaics and ecovoltaics. The aim of this book is therefore not to dangle unrealistic objectives but, by showing the avenues currently being explored around photovoltaics, to encourage the reader to reflect on the even more general prospects offered by multidisciplinary engineering that could respond to both climate and biodiversity issues.

Yvon LE MAHO
CNRS Emeritus Research Director
Institut pluridisciplinaire Hubert-Curien
University of Strasbourg and Scientific Center of Monaco
Member of the *Académie des sciences*
Associate Member of the *Académie nationale de pharmacie*

Foreword by Thomas Lesueur

In response to the two major contemporary challenges of combating climate change and preserving biodiversity, this book sheds light on two innovative and promising solutions: agrivoltaics and ecovoltaics, that is, the combination, on the same soil, of crops for food or ecological purposes and photovoltaic panels in conditions that are sustainable and even regenerative for the planet.

The following pages provide benchmarks, lessons from experience and perspectives on these techniques, which are still in the making, combining local solar electricity production and biodiversity-friendly agronomic practices on a single plot of land.

The main studies and summaries presented here on the subject – notably those carried out by ADEME (French Environment and Energy Management Agency) – remind us that the photovoltaic industry can still make progress in reducing its ecological footprint. Three particularly sensitive points are brought to the reader's attention: the conditions for extracting, processing and transporting silicon, the materials needed to build the installations and the recyclability of the material at the end of its life.

Second, lessons from experience: agrivoltaics and ecovoltaics have been tested in recent years on very different scales and in very different ways. The overview of initiatives carried out in France and abroad integrating these agro- and eco-energy approaches shows one of the major difficulties hindering their deployment: the lack of natural light for the crops protected by the photovoltaic panels.

Lastly, and most importantly, in response to these difficulties, solutions are outlined and tested in the field, in particular the cultivation of plants chosen specifically for their tolerance of low soil moisture and their food, medicinal, phytosanitary and ecological value. The first results obtained for these crops are encouraging; they are intended to provide food for thought on the future of our agricultural models.

Thomas LESUEUR
General Commissioner for Sustainable Development
Ministry of Ecological Transition, France

Introduction

The worrying figures are multiplying, and many indicators are in the red.

+ 0.53 °C: Average positive temperature anomaly in the Northern Hemisphere between 1997 and 2006, despite a decrease in solar activity.
+ 1.2° C: Temperature increase of tropical waters in the 20th century.
+ 1.0 °C: Increase in global air temperatures between 1850–1900 and 2018.
+ 0.63 °C: Temperature increase in May 2020 compared to the average temperature of May since the beginning of the data (1981).
+ 90%: Increase in global CO_2 emissions between 1970 and 2017.

Figure I.1. *Some numerical data illustrating global warming*

The multiplication of extreme climatic phenomena and anomalies (drought, intense heat waves, heavy rains, floods, etc.) is now recognized by all.

The growing consumption of and demand for energy has generated greenhouse gas waste with very noticeable consequences. According to Jancovici and Grandjean (2006), 80% of the energy consumed in the world contributes to global warming.

Faced with the climate emergency, new political decisions are gradually emerging.

The French law of August 17, 2015, on the energy transition for green growth was put in place to:

> establish a robust and sustainable energy model in the face of energy supply challenges, price trends, resource depletion and environmental protection imperatives (Legifrance, law no. 2015-992).

Specific targets have been proposed to reduce greenhouse gas emissions, reduce energy consumption and promote renewable energy.

Figure I.2. *The climate agreement, Paris, December 2015. The photograph shows Laurent Fabius, former Prime Minister of France and currently serving as President of the Constitutional Council since 2016*

More recently, the European Union has embarked on an ambitious climate policy. In November 2018, the European Commission presented a long-term strategy to achieve a climate-neutral economy by 2050, in order to implement the commitments made in the Paris Agreement (Actualité Parlement européen 2018).

This agreement was adopted at the Paris Climate Conference (COP21) in December 2015 (Commission européenne 2016). This was a historic event, as it represents the first global agreement on climate change. It was signed by 195 countries, including the European Union. As Laurent Fabius said, this is not a political agreement, but an agreement for the climate. It requires everyone to take responsibility.

The green pact or green deal for Europe proposes a roadmap that should lead to carbon neutrality in 2050 (Commission européenne 2019).

This goal is essential if we are to avoid disastrous climate change.

Energy transition, greenhouse gas reduction, carbon offsetting, carbon neutrality and the green deal are vital objectives. However, they are not the only ones. Everything cannot be reduced to a carbon rate. The energy transition must be carried out in an ecological manner, by giving ourselves the scientific and economic means capable of respecting life, whether human, animal or plant. It must be said that here too, many indicators are worrying:

– erosion of biodiversity, destruction of natural habitats of wild species, overexploitation of natural resources, multiplication of invasive animal and plant species that threaten already fragile ecosystems;

– sixth wave of extinction of species: a demonstrative example is the massive disappearance of insects, while the reproduction of 80% of flowering plants depends on them;

– environmental pollution: damage to aquatic systems threatening the quality of water resources, degradation of soil quality leading to a reduction in arable land at a time when demographic pressure is galloping and a reduction in air quality accompanied by the emission of particles, volatile organic compounds and gases harmful to humans, animals and plant species.

These few facts clearly prove that the energy transition must be ecological and environmental. All these problems are closely linked and intertwined.

How can we combine energy and ecological transition?

This book proposes to demonstrate that the implementation of renewable energies can be accompanied by an agricultural transition; that it can also accelerate ambitious and innovative ecological programs in harmonious interaction with many fields of activity (biodynamic agriculture, health ecology, alternative medicine, preservation of insect biodiversity and enrichment of the quality of poor and not very fertile soils, etc.). This transition is in line with the Sustainable Development Goals (SDGs), notably those responding to the global challenges of clean and affordable energy (SDG 7), the fight against climate change (SDG 13) and the preservation of life on Earth (SDG 15).

Figure I.3. *A necessary energy and ecological transition*

Among the different renewable energies (solar, wind, hydraulic, biomass (plants, organic waste), geothermal, marine), solar energy is the most advanced.

The sun, a natural and renewable source of energy, can lead to two uses:

– electricity production (photovoltaic solar energy);

– heat production (solar thermal energy) (Ministère de la Transition écologique 2020a).

In this book, we will focus on photovoltaic solar energy, which transforms solar radiation into electricity through photovoltaic cells integrated into photovoltaic panels. We will show how the development of photovoltaic parks can be integrated with new agricultural activities (*agrivoltaics*) and ecological activities (*ecovoltaics*).

This book is divided into three main chapters:

– *Chapter 1* introduces the basic concepts of photovoltaics and the photovoltaic industry through its various uses on rooftops, brownfield lands and drylands. It also addresses the degree of maturity of the industry and the thorny issue of life cycle assessment (LCA);

– *Chapter 2* deals with the cohabitation of photovoltaic and agricultural activities through agrivoltaics. The issues of cohabitation is addressed from different angles. A review of different experiences combining photovoltaics and agricultural activities is presented;

– *Chapter 3* presents recent initiatives that seek to associate a positive combination of photovoltaic parks and active ecology. This original approach, called *ecovoltaics*, is discussed through the first feedback from the field.

Figure I.4. *Structure of the book*

1

Photovoltaics: Concepts and Challenges

1.1. Brief description of the different photovoltaic cell technologies

Edmond Becquerel (Figure 1.1) was a French physicist who discovered the voltaic effect. He was the son of the physicist Antoine Becquerel and the father of Henri Becquerel, who discovered radioactivity with Pierre and Marie Curie. Edmond Becquerel demonstrated for the first time that certain materials could produce small amounts of electricity when exposed to light. A photovoltaic cell is composed of two silicon-based semiconductor layers.

Figure 1.1. *Photograph of Edmond Becquerel by Nadar*

The first layer is made of silicon, which is enriched by traces of phosphorus. The phosphorus atom has one more external electron than the silicon atom. This fifth additional electron can thus circulate easily between the atoms of Si (in blue) and P (in orange). A semiconductor material known as being of type N is thus formed (negative charges in excess) (Figure 1.2).

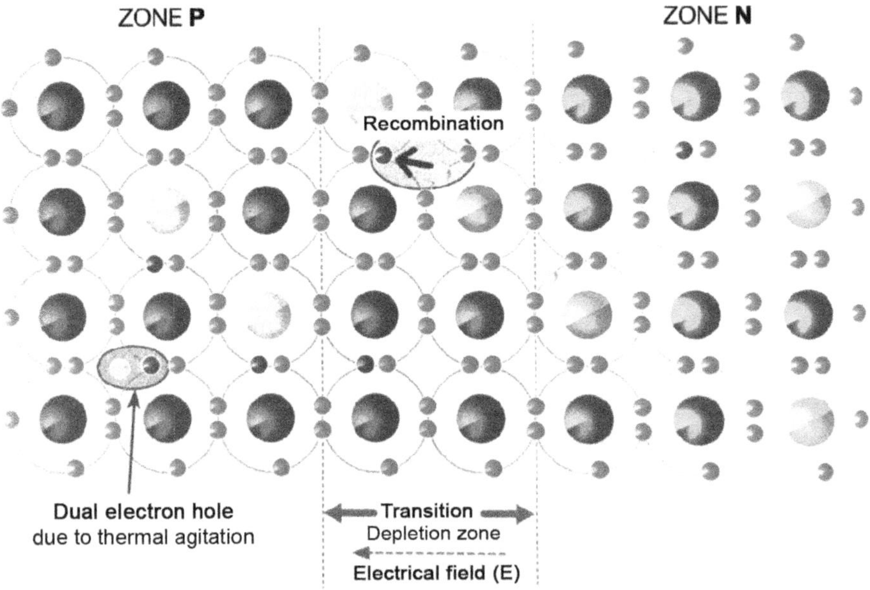

Figure 1.2. *Photovoltaic effect (source: Moine (2016)). For a color version of this figure, see www.iste.co.uk/grison/ecology.zip*

The second layer is made of silicon doped with traces of boron (or aluminum, in green). This time, boron (or aluminum) has one external electron less than silicon. The global atomic network thus has some electronic gaps, assimilated with a positive charge. A material known as type P is thus formed (positive charges in excess).

Following the shock of photons from the sunlight, the excess electrons (N layer) enter the P zone. This movement of electrons creates an electric field. By connecting the two layers with an electric circuit, the current can circulate (Figure 1.2). Thus, under the effect of solar radiation, the energy of the photons is transmitted to the electrons, which then transform it into electric energy.

The semiconductor materials are generally silicon and cadmium telluride. The first type is the most used. Silicon can be amorphous, semi-crystalline or crystalline depending on its mode of preparation, giving different color tones to the photovoltaic cells. The photovoltaic cells are then, respectively, gray, blue or blue and dotted with patterns left by the crystals. The structure of the semiconductor directly affects the performance of the voltaic cell and ultimately that of its assembly in solar panels. A comparative summary of the main technologies is presented in Table 1.1.

In France, the government finances up to 43% of the research and development of photovoltaic panels. Numerous studies are carried out to improve the yield of solar panels, using increasingly efficient technologies. This is one of the major challenges that academic research is currently facing in the field of photovoltaic energy. To achieve this, many efforts are being made in the production of new semiconductor materials, new cells and modules.

ADEME (*Agence de la transition écologique*, French Environment and Energy Management Agency) has defined several areas of research that are priorities in the strategic roadmap of photovoltaic solar energy in order to achieve the objectives set by the legislation. These areas of development for the photovoltaic industry are:

– technology/innovation/industry;

– building integration;

– services and network optimization;

– interdisciplinary themes.

The last area reflects the desire to develop a sector that generates an acceptable environmental footprint. This is a key issue, which reminds us that new technologies must be developed taking into account life cycle analysis. For example, high-performance CIGS[1] technology requires the use of cadmium sulfide. Cadmium is a highly toxic metallic element, regulated by REACH (Registration, Evaluation and Authorization of Chemicals). Research is underway to replace cadmium with less toxic compounds (Zn, Mg, O, S) or indium sulfide (a rare metal).

---

1 Abbreviation for copper, indium, gallium and selenium.

Technology	Description	Performance	Maturity
Amorphous silicon cell	The amorphous silicon photovoltaic cell is composed of a thin layer of silicon, much thinner than the monocrystalline or polycrystalline silicon cells.	7%	With poor performance, it is rather intended for solar watches, garden lighting and solar calculators
Monocrystalline silicon cell (1st generation photovoltaics)	This is the historical sector of photovoltaics. Monocrystalline cells are the first generation of solar cells. They are made from a block of silicon crystallized in a single piece.	16–20%	35% of the global market
Polycrystalline silicon cell (1st generation photovoltaics)	Polycrystalline cells are made from a block of silicon composed of multiple crystals. They are often found in domestic, agricultural and industrial installations.	Approximately 15%	55% of the global market (best value for money)
Thin layers (2nd generation photovoltaics)	The cells are composed of layers of semiconductor and photosensitive materials on a support made of glass, plastic, steel, etc. Different materials can be used, the most common being amorphous silicon.	5–15%	10% of the global market Decrease in production cost
Organic cells (2nd generation photovoltaics)	The use of polymeric materials aims to replace mineral materials with organic semiconductors, that is, plastics, for the manufacture of photovoltaic cells. These are cheap, have good absorption properties, are easy to deposit and are much lighter than the 1st and 2nd generation materials.	20–30%	In demonstration

Technology	Description	Performance	Maturity
Concentration cell (CPV technology)	This technology uses optical lenses that focus light onto small, high-performance photovoltaic cells. It is necessary to always be positioned facing the sun (mobile system).	5–10%	Experimental stage
Cadmium telluride cells	The manufacturing process hermetically seals a single cadmium telluride absorption layer between two glass supports.	10%	–
CIGS cells	The semiconductor material is an alloy of copper, indium, selenium and gallium. This mixture is arranged in a very thin layer on a support. In high-performance technologies, the CIGS structure forms a complex junction made of materials of different natures (heterojunction) of the CIGS(p)/CdS(n)/ZnO(n) type.	13–23%	> 10%
Hybrid perovskite cells	Perovskites are crystals formed by molecules containing an organic and an inorganic part. They are emerging as a new class of semiconductors that show exceptional performance (Deleporte 2017).	22%	Research stage

Table 1.1. *Different technologies of photovoltaic cells*

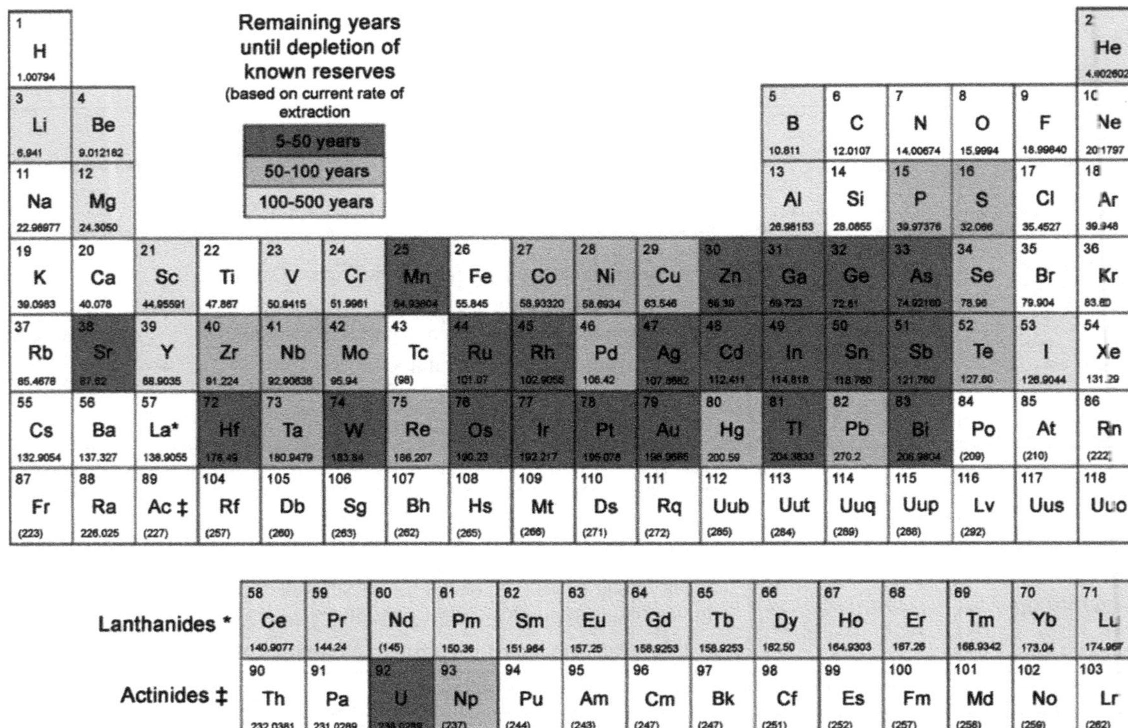

Figure 1.3. *Global depletion of mineral resources (Hunt* et al. *2015). For a color version of this figure, see www.iste.co.uk/grison/ecology.zip*

The same is true for cadmium telluride technology, the nanoparticles of which are highly toxic. In the case of release of the Cd^{2+} cation, genotoxic, reprotoxic and endocrine-disrupting effects may arise (Cadmium 2010).

The elements zinc, indium and gallium pose the problem of the depletion of mineral resources. A review published in *Green Chemistry* (Hunt *et al.* 2015) confirms this major problem for the photovoltaic sector in a short-term shortage of widely used metal species.

This problem echoes Guillaume Pitron's book (Pitron 2018). Of particular interest is an investigation by the director of the Canadian company Uragold, Bernard Tourbillon, that produces the materials needed for the solar industry. A detailed calculation of the ecological impact of photovoltaic panels suggests that the production of a silicon-based solar panel generates 70 kg of CO_2. Assuming that the number of photovoltaic panels will increase by 23% per year in the coming years, solar installations will produce 10 gigawatts (10 GW) of additional electricity per year, and generate 2.7 billion tons of CO_2, equivalent to the pollution generated by the annual activity of 600,000 automobiles (Pitron 2018).

These data prove that photovoltaics must be thought through in terms of life cycle analysis (LCA): from the design of semiconductor materials to the recycling of these same materials after use. Photovoltaic industry players must adopt a dual LCA and circular economy approach and integrate it into their development.

This is the only way that photovoltaics can be considered as green energy and meet the great energy challenge, which, let us remember, is also ecological. This consistency of objectives is important, especially since France has real potential in solar energy.

The sunshine statistics of the French territory are indeed favorable for the development of solar energy (Figure 1.4) (Kalyanpur *et al.* 2013). Solar power plants are therefore a possible choice to transform the French electrical system (Ministry of Ecological and Solidarity Transition 2020).

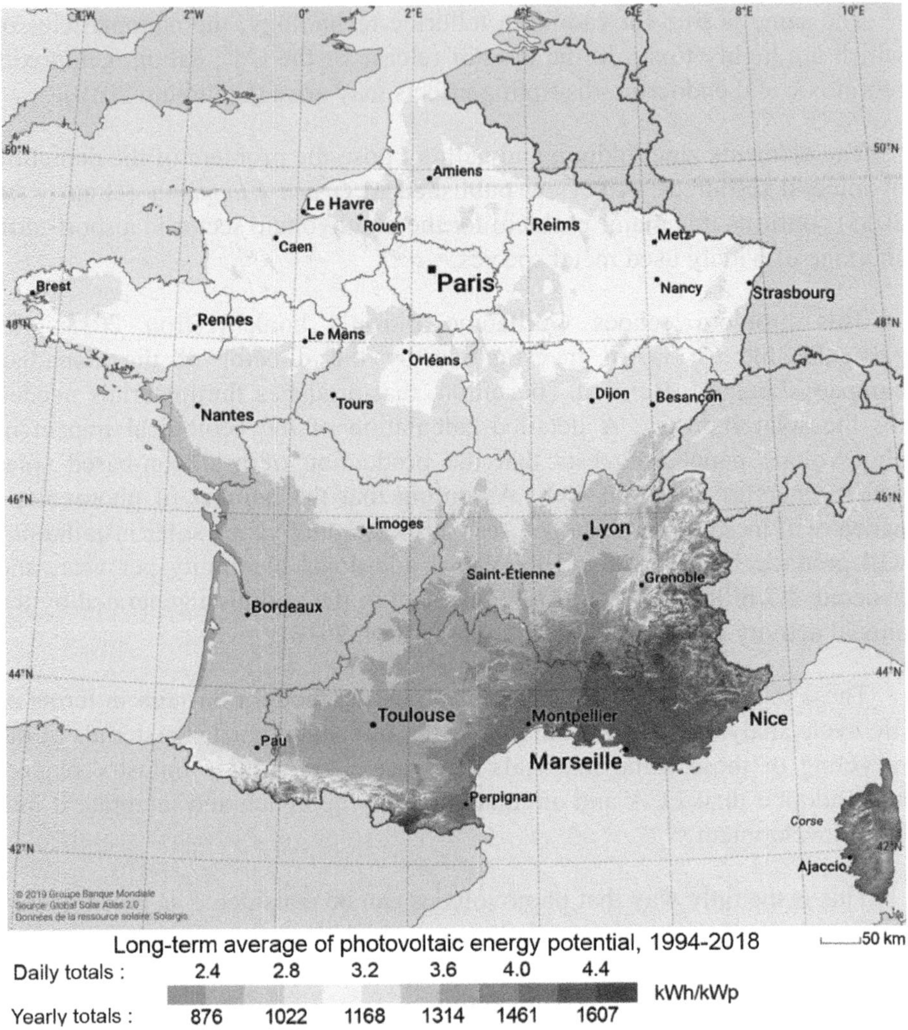

Figure 1.4. *French solar energy potential (source: Global Solar Atlas 2.0 © 2020 The World). For a color version of this figure, see www.iste.co.uk/grison/ecology.zip*

Electricity is one of the three pillars of energy consumption, along with transportation and heat production. It represents about 22% of final energy consumption in France (Ministry of Ecology, Energy, Sustainable Development and the Sea 2011).

In addition, photovoltaic electricity has seen a steady decline in production costs and is now a non-negligible source of renewable energy, thanks to accessible technology (ADEME 2017).

1.1.1. *Photovoltaic versus thermal solar panels*

Solar energy can be used for two different purposes:

– heat production;

– electricity production.

A solar thermal system provides hot water for domestic use or for heating. A photovoltaic installation produces electricity. This electricity production can either be used directly by an electronic device such as a watch or be connected to the electrical network.

The choice of uses is a real societal responsibility. Solar thermal energy can be a highly water-consuming technology, especially in the case of solar farms and power plants installed in arid areas (up to 3,500 liters per megawatt-hour (MWh) (Pitron 2018)).

1.2. Different types of photovoltaic installations

There are two main types of electricity production using photovoltaic panels connected to the electrical grid. The first one concerns the installation of solar panels on the roof of an energy-consuming building. The use can be for domestic purposes in private homes (individual houses, social or private housing) or for professional purposes (offices, shops, public facilities, industry, agriculture).

The second type of photovoltaic installation is the ground installation. This installation is made up of numerous modules supported by structures, whose production directly supplies the electricity network.

At the end of 2018, photovoltaics represented 9 GW in France.

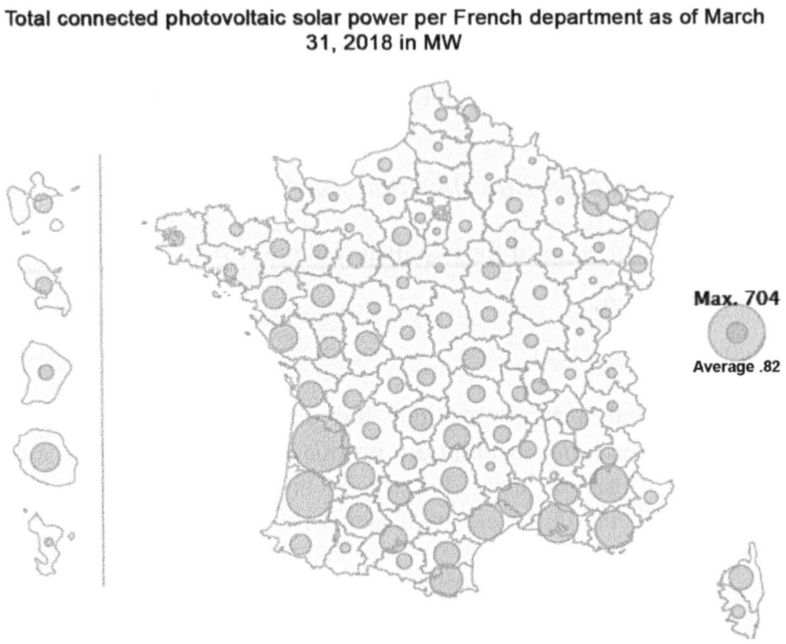

Figure 1.5. *Photovoltaic power in France, 2018. (source: Service des données et études statistiques (SDES), 2018 survey). For a color version of this figure, see www.iste.co.uk/grison/ecology.zip*

1.3. Legislation

In addition to the Climate Plan, the Grenelle law sets the share of renewable energy in French consumption. Other more specific laws have been drawn up to support the development of renewable energies such as photovoltaics.

The Multiannual Energy Program (*Programmation pluriannuelle de l'énergie*, PPE) was published in the *Journal officiel* (JO) on April 23, 2020. It provides for a fourfold increase in grid-connected photovoltaic capacity in 2028 compared to today. This decree sets a low target of 35.1 GW of photovoltaic capacity installed in France in 2028 and a high target of 44 GW (Decree No. 2020-456 of April 21, 2020, on the multi-year energy plan).

Article L111-18-1 of the urban planning code, amended by the Climate and Energy Law of November 8, 2019, provides that building permits for buildings with a footprint greater than 1,000 m^2 will have to incorporate a renewable energy production process or a vegetation system. Photovoltaics can be a response to this requirement. If the building permit also includes a parking lot, some or all of the photovoltaic system can be installed on shaded areas of parking lots. The renewable energy production process or the greening system must then cover at least 30% of the total surface area of the roofs and shade created.

In addition to these development objectives, other regulations are in force for the carrying out of photovoltaic installations, whether on the roof or on the ground. These procedures depend on the size and the location of the installation.

Before addressing the combination of uses (agriculture/photovoltaics; ecology/photovoltaics), it is appropriate to assess the photovoltaic solution by outlining the energy, economic, environmental and ecological trade-offs.

1.4. Advantages of photovoltaics

Solar energy is, on a human scale, inexhaustible and freely available in very large quantities.

Photovoltaic panels can be used at different scales, and their installation is easily scalable and adaptable to the conditions of the place (on a building or on the ground).

Photovoltaic technology is more modern than nuclear and wind power; recycling is structured.

The acquisition of this mode of energy can be individualized and become a citizen choice.

A photovoltaic installation does not generate greenhouse gases during its operation. It produces little hazardous waste and does not emit local pollutants during its operation. The International Energy Agency calculates that a photovoltaic installation connected to the grid provides the equivalent of the energy needed to manufacture it within one to three years (the thin-layer technology is more quickly compensated than the silicon technology) depending on the amount of sunshine on the site.

1.5. Disadvantages of photovoltaics

Due to the irregularity of sunshine between summer and winter, the choice of the sunshine period used in the power calculation depends on the way the photovoltaic energy is used. Large-scale renewable energies require an adaptation of the electrical system and the networks because of the intermittency of production.

The location of the site determines several criteria for the installation of photovoltaic panels, such as their inclination to the Sun (for the system to be optimal, the incident beam of the Sun must be perpendicular to the surface of the panel) and the installation methods according to geographical constraints (location of the site as well as topology).

Electrical energy cannot be stored "directly", that is, in its primary form. However, it is possible to store it "indirectly" in batteries in chemical form or in kinetic accumulators in mechanical form.

Photovoltaics do not produce electricity at the end of the day when consumption is the highest.

Photovoltaic technology requires large surfaces because the solar radiation captured is proportional to the exposed surface.

Figure 1.6. *Photovoltaics and solar radiation capture surfaces (source: C. Grison)*

1.6. Some figures on the environmental footprint compared to other energy sources

The IPCC (Intergovernmental Panel on Climate Change) has assessed the production of greenhouse gases based on the amount of electricity produced (GIEC-IPCC 2014). While coal, followed by gas, are by far the most problematic, solar voltaic technology is not the best placed energy source (as shown in Table 1.2).

Coal	Gas	Biomass	Other solar PVs	Solar PV roofs	Geothermal energy	Solar concentration	Hydraulics	Nuclear	Marine wind	Onshore wind
820	490	230	48	41	38	27	24	12	12	11

Table 1.2. *Greenhouse gas emissions by form of electricity generation, average values in gCO_2 eq/kWh (Cassoret 2020)*

Nature of the pollutant	Coal	Gas	Biomass	Photovoltaics	Hydraulics	Nuclear	Wind
SO_2	32,321	15,000	140	490	60	50	87
NO_X	5,273	1,500	1,950	340	42	100	50
VOCS	29	164	0	70	0	0	0
Fine particles	663	10	320	190	5	2	35

Table 1.3. *Maximum emissions of sulfur dioxide, nitrogen oxides, non-volatile organic compounds and fine particles generated from the production of 1 kWh of electricity, in mg per kWh (Cassoret 2020)*

The same applies to the emission of sulfur dioxide, nitrogen oxides, non-volatile organic compounds and fine particles generated if we reason in terms of life cycle analysis, that is, taking into account the extraction of raw materials up to the waste produced.

The production costs of photovoltaics are higher than those of non-renewable resources and nuclear energy (Figure 1.7). However, these figures reflect confusion about the assessment of the carbon impact of nuclear energy over its entire life cycle, and this conclusion can be wiped away (Crampon 2019). These inaccuracies are reinforced by the lack of feedback on nuclear power plant decommissioning operations in France (National Assembly 2020).

In 2015, Hertwich *et al.* published a very fine-grained life cycle analysis of green energy pathways.

The findings indicate that the large-scale implementation of photovoltaics and wind power has the potential to reduce pollution and other environmental impacts of electricity generation, such as greenhouse gas emissions, freshwater ecotoxicity, eutrophication and particulate exposure. However, the construction of the installations results in high consumption of aluminum and copper. Photovoltaic energy requires up to 40 times more copper than the combustion of fossil fuels. Given the high consumption of copper, this metal element could become a concern. However, the pollution caused by these materials is still lower than that caused by direct emissions from fossil fuel power plants.

The production cost of photovoltaic solar energy depends in part on the available sunshine and therefore on the productivity of the solar panels. In France, the potentially producible photovoltaic energy varies between the north and the south of France. On average, the expected productivities in the north and south of France are about 950 kWh/year and 1,400 kWh/year. Therefore, the production costs of electricity depend on the location of the solar panels. It would be interesting to study the line losses due to the distance between the production and consumption locations, which is greater for large photovoltaic plants than for small installations.

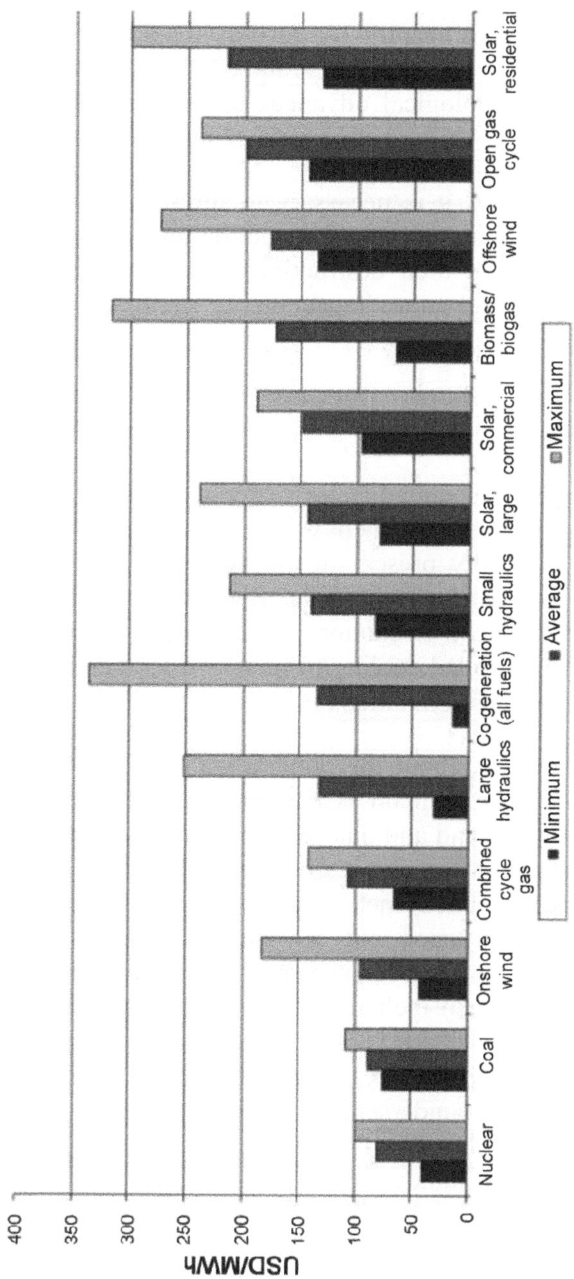

Figure 1.7. *Cost of electricity production in 2020 (source: DGEC, Ministère de la Transition écologique). For a color version of this figure, see www.iste.co.uk/grison/ecology.zip*

The investment costs of the sector depend on the type of photovoltaic panels and their installation. The investment costs of photovoltaic solar energy have decreased significantly for several years and are expected to decrease further, by 30% between 2015 and 2025 according to ADEME estimates, thanks to technological advances and the increase in installed capacity (ADEME 2018a). Utilities fund 60% of solar power projects.

As has been mentioned, it is necessary to think in terms of life cycle analysis (from the production of materials and the manufacturing and use of photovoltaic installations to their destruction) of a technology, in order to better understand its real and future impact on our environment.

1.7. Origin of the silicon needed for the construction of photovoltaic cells

As such, the origin of semiconductor materials is a key point. Most of them are based on silicon. This element is very abundant in the Earth's crust. However, it is not naturally present in the degree of oxidation useful in photovoltaic cells, nor in the right state. Significant transformations are necessary: amorphous silicon is obtained from silicon gas, monocrystalline silicon is formed by controlled cooling of melted silicon and polycrystalline silicon results from the fusion of silicon in an ingot mould.

A few technical details about the preparation of polycrystalline silicon reflect the effort involved. Silicon is derived from silica, silicon dioxide (SiO_2), which is found in sand and quartz. The refining of silicon consists of separating these atoms in a furnace, a step that consumes a lot of energy. It leads to liquid silicon, called metallurgical silicon. The purification of metallurgical silicon is carried out by chemical processes to provide polycrystalline silicon. For its use in the photovoltaic industry, silicon must be further purified. It is then melted and recrystallized into ingots used to manufacture photovoltaic cells. It is at this stage that the dopants needed for the cells to function properly are added (Belkhayat 2017). This quick analysis provides a better understanding, initially, of the calculations of B. Tourillon and the analysis of G. Pitron mentioned above. The preparation of silicon needed to build photovoltaic cells is energy- and input-intensive. However, innovations in silicon extraction and purification techniques have led to significant improvements. In particular, the thin-layer technology has a significant degree of maturity (Boxenbaum *et al.* 2013). Finally, third-generation photovoltaic cells, made of organic molecules, may be a solution, if they are synthesized in an environmentally friendly manner.

1.8. End of life of solar panels

All manufacturers today offer 25 year production guarantees (production is 90% of the initial production after 10 years and 80% after 25 years). Existing installations show that the modules can produce for 30 years. At the end of the installation's life, the operator has two choices:

– continuing the activity, which requires the replacement of the production modules by new-generation modules and the modernization of the annex installations (subject to obtaining new administrative authorizations and the renewal of the lease of the land);

– halting the activity, which requires the deconstruction of the installations, the collection of materials and the restoration of the site.

1.9. Degree of maturity of material recycling

The recycling of solar panels has developed considerably in recent years. Recycling channels for photovoltaic panels have been organized in France and in Europe, especially since the creation in 2007 of the PV Cycle association, which organizes the collection and recycling of French and European photovoltaic panels. A crystalline silicon-based photovoltaic module with an aluminum frame is now 94.7% recoverable.

Since 2014, manufacturers and importers of photovoltaic panels have been legally obliged to take back end-of-life solar equipment free of charge and to contribute financially to the collection and processing of the waste. Most manufacturers of photovoltaic panels have a share of their turnover that is paid to PV Cycle. These efforts are the best solution to limit the need for raw materials and avoid dependence on China (Greenpeace France n.d.).

In conclusion, photovoltaics are an example of successful waste recycling in the field of green electricity.

1.10. Location and mode of development of photovoltaics

1.10.1. *Photovoltaic panels on roofs*

Most often, photovoltaic installations on a building are carried out on a south-facing, unobstructed roof. The range of powers concerned can be very large depending on the available surface, going from a few square meters to several hectares.

The installations can be "building-integrated" (the photovoltaic system is installed in the plane of the roof of which it ensures the waterproofing) or in the overlay of a roof (the photovoltaic system is installed on top of the roof).

The processes of integration into the building are very diverse: it is possible to equip a flat or sloping roof with different products depending on the element of the roof (tiles or slates, metal tanks, waterproofing, etc.) or the facade (cladding, railings, sunshades, etc.) considered. When integrated into the building, the solar panels also provide waterproofing and resistance to mechanical loads. In addition to the essential functions of the building, aesthetic functions are sometimes deployed. This integration of photovoltaic systems reduces the costs of the substituted elements. However, it must be taken into account that the system will have to be replaced or renewed at the end of its life, which could present significant costs.

The ADEME evaluates the capacity of the photovoltaic deposit on roofs in France at a height of 364 GWp (gigawatt-peak), including a development deposit of 120 GWp on large roofs (ADEME 2018b).

1.10.1.1. *In the industry*

The industry sector is the third-largest French energy consumer, accounting for 19% of final energy consumed in 2014, after the transport and residential sectors.

Since the law on energy transition for green growth (LTECV) aims to increase the share of renewable energy to 32% of final energy consumption in 2030 (compared to 14.6% today), developing the integration of renewable energy in industry is necessary to achieve this goal.

According to ADEME, the integration of renewable energy in industry is distinguished by three criteria:

– low-temperature energy production: $T < 150°C$;

– the production of high-temperature energy: $T > 150°C$;

– the production of electrical energy.

Photovoltaic solar energy is more suitable for the production of electricity, which can supply any sector with varying degrees of efficiency. This sector, due to its maturity and availability, is perfectly suited to industry. However, it is still not very developed for self-consumption in industry, since historically photovoltaic solar energy has been used more for injection into the electricity grid. The French energy mix (i.e. the energy sources used in the production of electricity in France) already has a low carbon impact. Therefore, the reduction of the carbon footprint brought by the photovoltaic sector has not been clearly demonstrated.

However, there are some points to consider when installing solar panels on industrial buildings. For example, to install solar panels on a roof, the roof structure should support at least a weight of 15 kg/m^2 (including the panel structure). In some cases, it may be necessary to reinforce the roof.

The sector is experiencing an overall easing of administrative procedures and constraints. Some types of projects no longer require a building permit, such as large-scale PV roofs on existing buildings for self-consumption (see ADEME (2018a)).

The Atrya Group, which includes Tryba Industrie, specializes in the manufacture of windows, doors and shutters. The group is present in France and in Europe and has 21 production sites. The many industrial processes involved in the manufacture of windows and doors, such as machining, painting, assembly, glazing and storage, are very energy-intensive. The site's electricity consumption is significant, reaching up to 3 GWh per year.

Tryba Industrie has decided to implement a self-consumption photovoltaic power plant. The objectives of this project are to promote a local and green energy and to reduce its energy footprint by becoming an actor in its own energy consumption. This project is now the largest self-consumption photovoltaic power plant in the Grand-Est region. It was the winner of a call for proposals launched by the Grand-Est region and ADEME in 2015 to support photovoltaic self-consumption projects.

The investment for the photovoltaic installation was €320,000 (excluding VAT). It was subsidized up to 30% by the Grand-Est region as part of its 2015 call for self-consumption photovoltaic projects. The estimated payback time for this installation is between 8 and 10 years, for an average solar panel life span of about 30 years.

The self-consumption photovoltaic plant covers 5% of the site's annual electrical needs, which is equivalent to the annual electrical consumption of 140 households.

The agricultural cooperative Alp Coop (Laragne, France) sorts, packages and stores about 20,000 tons of fruit per year in a refrigerated environment. The cooperative has installed 1,200 photovoltaic solar panels for a total power of 300 (kilowatt-peak). Part of the electricity produced is used to power the cold storage rooms, which represent 30% of the site's needs. The rest of the production is fed into the grid. The installation cost €490,000 and benefited from ADEME grants as part of the 2016 call for self-consumption projects.

Box 1.1. *Examples (ADEME 2018b, 2018c)*

1.10.1.2. *In agriculture*

The final energy consumption of agriculture represents 3% of the total energy consumption in France. Electricity is the second most consumed energy source. It is mainly used for heating and ventilation of livestock buildings.

As for other sectors of activity, one of the challenges of agriculture is to succeed in limiting energy consumption. One solution may be to reduce energy requirements by building increasingly optimized and adapted infrastructures. One of ADEME's recommendations is to invest in renewable energy (ADEME 2015).

One solution could be to insert energy production in the construction of buildings, while respecting the recommended zootechnical bases: it is a compromise between livestock or storage buildings and electrical production.

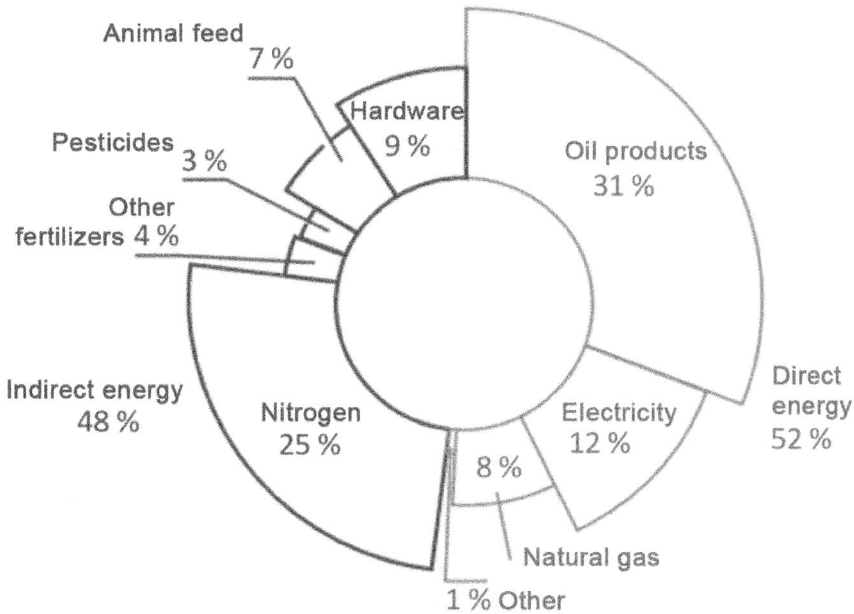

Figure 1.8. *Breakdown of direct and indirect primary energy consumption in agriculture (source: ADEME, 2013). For a color version of this figure, see www.iste.co.uk/grison/ecology.zip*

According to the *Ministère de la Transition écologique et solidaire* (French Ministry of Ecological Transition and Solidarity), the first operation to install photovoltaic panels in agricultural facilities was launched in 2008 and led 110 farmers to join together to build a photovoltaic park spread over 190 agricultural buildings and representing 7 MW of installed power. In 2013, a second operation led to the installation of 140 power plants of 9 kW (60 m^2 per building). Two hundred projects of this type are currently being investigated for a third collective operation, which this time will call on citizen savings to complete government funding (Ministère de la Transition écologique et solidaire 2017).

The advantages of photovoltaic structures in agricultural buildings are the large usable roof area, usually with large building areas, and the creation of an interesting additional income for an activity sometimes economically fragile or threatened by land pressure.

Figure 1.9. *Agricultural shed with photovoltaic roof in the vineyards of Jurançon (Domaine de Bellegarde) (source: P. Labasse)*

The disadvantages are the distance from the network when connecting the installations in natural areas and the sensitive landscape context. A standardized generalization of this assembly, without any architectural or landscape insertion approach, could lead to a banalization and a depreciation of the landscapes.

However, it must be considered that there are points of vigilance to be controlled, such as the establishment of greenhouses or sheds whose use would be more to produce electricity for sale than to develop the agricultural activity (Direction départementale des territoires et de la mer de l'Hérault 2011). Another important point has been made by *France Nature Environnement* (FNE). In order to be truly ecological, photovoltaic projects must not in any way aggravate the artificialization of natural and agricultural areas. Meeting electricity needs must not be done at the expense of already threatened biodiversity. For example, a vast photovoltaic park project is being studied in the Lot-et-Garonne region. Solar panels are to be spread over five communes, on nine "farms", and cover 1,300 hectares of farmland

and 700 hectares of forest. It is true that the compensation law stipulates that for every hectare of forest cut down, at least two others must be planted. However, this strategy is contrary to stopping the artificialization of land and the erosion of biodiversity.

Faced with an unprecedented ecological crisis, covering and equipping already artificialized areas should be preferred, in order to preserve the diversity of plant and animal species.

In a study published on May 24, 2019, ADEME (2019a) confirmed that France has considerable potential for the installation of solar panels in locations unlikely to give rise to conflicts of use (see section 1.10.2.1.3).

1.10.1.3. *Public spaces*

Strengthening the consideration of sustainable development in projects is a major concern in public spaces.

Public buildings, secondary and tertiary activity buildings, have the advantage of being able to accommodate photovoltaic panel installations on the roof or in large cladding. The potential roofs and facades that can be equipped are significant. This situation is often found in dedicated areas away from residential areas, with generally limited landscape and social acceptability constraints, and surfaces that are generally large and in one piece (Direction départementale des territoires et de la mer de l'Hérault 2010).

Original initiatives have been carried out by Montpellier Méditerranée Métropole. The business centers are equipped with solar facades, a shade appreciated during the long sunny months.

Other similar initiatives are multiplying. For example, the company ORA and the city of Montpellier are installing self-consuming parking lot shades for the municipal swimming pool in Cournonterral (Hérault), which will account for 30% of the pool's annual electricity consumption. In the Eure region, many nurseries and elementary schools have roofs equipped with solar panels. Some wooden shelters in school yards are equipped with transparent photovoltaic panels.

Parking lots shaded by a photovoltaic roof are coupled with charging stations for electric cars. Numerous initiatives are being put in place. Their development is facilitated by the 2019 finance law, which allows local authorities whose public buildings are equipped with photovoltaic installations to be exempted from property tax on the sale of electricity.

Figure 1.10. *Solar facade of a Montpellier Méditerranée Métropole business center (source: C. Grison)*

1.10.1.4. *Individuals*

Residential photovoltaics represents installations of up to 9 kWp on the roofs of houses for small installations and up to 36 kWp for medium installations.

The potential production of photovoltaics in France is 240 GWp on residential roofs (ADEME 2018b).

For private individuals, there are three possibilities when producing energy. First, the energy is totally injected into the network, that is, the owner of the electricity production installation sells the totality of their electrical production. For this type of installation, the dimensioning consists of producing the most energy possible according to the available surface and the investment capacity of the owner.

The electrical production can also be consumed on site and the surplus of production sold and injected into the electrical network. This is partial self-consumption.

The last option is to directly consume the totality of the produced energy, that is, total self-consumption. The development of self-consumption is encouraged by the ADEME in order to control the energy expenditure of individuals.

Residential self-consumption in France is growing more and more, as 8,000 applications were recorded for the year 2016 and 6,000 for the first half of 2017; 84% of the installations carried out in 2018 were self-consumption, 95% of which did not have a storage system and injected the surplus into the electricity grid (Observatoire des énergies renouvelables 2019).

For a stand-alone installation, the best compromise must be found between the electricity needs and the investment cost.

ADEME anticipates that by 2050, applications using photovoltaic systems will have diversified significantly. They will also concern residential, tertiary, industrial and agricultural activities, with a view to optimizing energy systems. Four areas of application have been selected:

– decentralized sites, where photovoltaic systems are installed on the roofs of individual houses, office buildings, sheds or greenhouses. The electricity produced could be consumed on site, stored or fed back into the grid with a value that can cumulate the sale of kilowatt-hours and the purchase and sale of services to the grid. This could be done through electricity sellers that aggregate the output of these thousands of dispersed units. In the future, products and applications outside of buildings will develop and diversify the photovoltaic market (street furniture, roads, deployable systems that can be used only outside the growing season, etc.);

– new buildings (housing, tertiary) with zero energy or positive energy;

– solar districts that combine housing, industrial and commercial activities, where optimal energy management is a challenge for the years to come. By 2050, the housing stock should be composed of 36 million dwellings divided as follows: an "old renovated" stock of about 27 million dwellings and a "new" stock of about 9 million dwellings;

– ground-mounted power plants that produce electricity sold on the electricity market or used locally by industrialists concerned about their supply.

1.10.2. *Ground-mounted photovoltaic panels: creating solar fields on unused industrial wastelands and parking lots*

Ground-mounted photovoltaic systems consist of rows of panels mounted on wooden or metal frames.

Ground-mounted photovoltaics can have two configurations:

– solar panels can be integrated into structures that do not consume electricity but for which the panels fulfill a well-identified function in addition to the production of electricity (parking lot shades, public passageway or station platform covers, noise barriers, etc.), which represent surfaces of several hundred to several thousand square meters;

– the second configuration is on bare land, also called "solar farms". These installations can have as sole purpose the production of electricity, and in this case, the surfaces used are from a few thousand to several tens of thousands of square meters.

For ground-mounted installations, favorable land is preferably flat and open, several hectares in size, easy to access and able to be fenced off (to protect the solar field, for insurance reasons). The distance of the connection to the nearest source station is also one of the criteria for selecting a site. The project must also be compatible with the local urban planning document and any applicable protections and easements (Centre national de ressources photovoltaïques 2020).

We can also distinguish two types of ground installations: fixed and mobile. The fixed installations are oriented to the south with an angle of exposure that can vary from 25° to 30° depending on the local topography. The mobile installations, called "trackers", are equipped with a motor allowing them to follow the course of the Sun in order to optimize their exposure and thus their output. They require a greater investment and maintenance but have a higher productivity.

1.10.2.1. *Legislation and characteristics*

The French government has put in place measures to regulate land use. Decree No. 2009-1414 of November 19, 2009, introduces specific provisions for ground-mounted photovoltaics in French town planning law. In particular, it extends the obligation to carry out an impact study and a public inquiry for ground-mounted installations of more than 250 kWp and also regulates the use of land by these plants.

The construction and operation of a ground-mounted photovoltaic power plant require approximately one to two hectares of land for 1 MW of capacity. This consumption of land is at the origin of a regulatory framework (local and national) aimed at respecting territorial planning rules, preserving biodiversity and landscapes and taking into account the human environment, particularly agricultural activities. This regulatory framework is notably enshrined in the Energy Code, the Urban Planning Code and the Environment Code.

Without constituting a mandatory regulatory framework for project developers, several planning instruments make it possible to set trajectories for the development of renewable energies in the territories, from the regional level (in particular with the Regional Schemes for Development, Sustainability and Territorial Equality (*schémas régionaux d'aménagement, de développement durable et d'égalité des territoires*, SRADDET)) to the level of the Public Establishment for Inter-municipal Cooperation (*Établissement public de coopération intercommunale*, EPCI) or the municipal level (Sustainable Development and Planning Project (*Projet d'aménagement et de développement durable*, PCAET)). These instruments also make it possible to define the “spirit” of the development of renewable energies in social, economic, landscape and biodiversity preservation terms.

1.10.2.1.1. The need for facilities to be consistent with the local urban planning document

At the local level, the local urban planning document is binding for project owners. This local urban planning document may take the form of a Land Use Plan (*Plan d'occupation des sols*, POS), a Local Urban Planning Scheme (*Plan local d'urbanisme*, PLU) or an Inter-municipal Local Urban Planning Scheme (*Plan local d'urbanisme intercommunal*, PLUi). In the absence of a local urban planning document, the National Urban Planning Regulations (*Règlement national d'urbanisme*, RNU) apply. In addition, the

Urban Planning Code defines the rules that apply to local urban planning documents in the mountains and on the coast, via the "Mountain" and "Coastal" laws.

The PLU regulations determine local zoning between natural and forest zones (N), agricultural zones (A), urban development zones (AU) and urban zones (U). For each type of zone, the PLU defines the type or nature of possible activities and the potential construction rules.

The construction of ground-mounted photovoltaic parks does not pose any difficulties from the point of view of compliance with town planning regulations in urban or urbanized areas. On the other hand, agricultural, natural and forest areas are considered as sectors to be preserved. In these areas, the installation of ground-mounted photovoltaic parks may be considered inappropriate.

In the N or A zones, the EPCI may decide to proceed with a procedure for changing the local urban planning rules to allow the implementation of a project. This procedure is highly regulated, with the obligation to carry out a public inquiry and the consultation of several services and commissions, in particular the *Commission départementale de préservation des espaces naturels, agricoles et forestiers* (CDPENAF, French Departmental Commission for the Preservations of Natural Agricultural and Forest Spaces). Finally, still in the N and A zones, investigating authorities can examine the authorization of photovoltaic projects if the facilities are considered to be of "public utility".

1.10.2.1.2. A mandatory individual authorization to build a ground-mounted photovoltaic power plant

Ground-mounted photovoltaic power plants generally have a capacity of more than 250 kWp and cannot be built before an environmental assessment (environmental impact study of the project and public inquiry) and the issuance of a building permit by the Prefect (French state's representative in a department or region).

In some specific cases, a clearing authorization, a derogation for protected species or an authorization under the water law are required.

The project's environmental impact assessment must (Ministère de l'Écologie, du Développement durable, des Transports et du Logement 2011):

– justify the choice of the site on a territorial scale;

– present the initial state of the site, the impacts of the project on biodiversity and landscapes, as well as the measures taken by the project owner to avoid, reduce or offset the impacts of the project (ARO logic);

– present the approach and methodology used for the impact study, as well as all the difficulties encountered.

The public inquiry is carried out by an investigating commissioner and is governed by the Environmental Code (Ministère de la Transition écologique et solidaire 2020b).

1.10.2.1.3. Issues

Ground-mounted photovoltaic plants are one of the most cost-competitive renewable energies. Ground-mounted installations have lower electricity production costs than rooftop installations (ADEME 2019a).

However, they also have a larger direct footprint than most other renewable energy technologies.

From the perspective of a strong development of photovoltaics on national territory, abandoned areas and parking lots offer surfaces with little or no competition of use: either the surfaces are no longer used (abandoned areas) or a photovoltaic installation can work in perfect cohabitation with the primary use of the site (parking lots with photovoltaic shades).

ADEME has already identified 17,764 sites as being suitable for the installation of photovoltaic plants. Among these sites, the most represented and identified in a study (ADEME 2019a) were former oil deposits (> 23%), former sites of commercial, craft and mechanical industry activities (garages, mechanical workshops, wrecking yards, forges, etc.) (> 5%) and former waste storage sites (> 4%).

The various derelict spaces can be classified into seven categories:

– old quarries and mines;

– old slag heaps;

– brownfield lands;

– commercial wastelands;

– waste storage sites;

– former military sites;

– parking lots.

Recommendations from ADEME are very clear:

> However, the development of photovoltaics must take into account environmental constraints such as land use. Panels must be installed on unused spaces such as large roofs (warehouses, commercial buildings) as a priority, and the implementation of ground-mounted plants must comply with strict environmental criteria (ADEME 2019b).

When creating a photovoltaic project, several criteria must be considered in order to see your project emerge and be implemented. These different criteria depend first of all on the regulations (environmental code, town planning, etc.), the rules of land use planning and the different obligations to be taken into account for the creation of the solar park (depending on the power of the park). Then, the specificities and constraints of each site must be studied and will influence the technical specifications related to the installation of solar panels. Depending on the site, it is also necessary to take into account the various types of pollution generated by the exploitation of these spaces and the stakes for biodiversity.

For example, the distance between the point of production of electricity and the point of connection to the grid is an important criterion for the realization of photovoltaic projects.

The interest of using abandoned spaces as listed here is that the tenders favor the use of already artificialized lands (urbanized, industrial or commercial areas, parking lots) over natural areas. Another interesting aspect of the installation of solar fields on this kind of space is the rehabilitation and valorization of the site.

This use also makes it possible to respect the challenges of sustainable development, in particular the limitation of land artificialization, the preservation of agricultural and natural land as well as landscapes. The achievement of photovoltaic development objectives in line with these issues is a major challenge (Ministère de la Transition écologique et solidaire 2020).

The identified potential deposit is 53 GWp. Despite the uncertainties related to the quantification of the deposit, the potential of the abandoned areas is important. Of the 53 GWp, 93% is distributed in neglected areas (49 GWp) and 7% in parking lots (4 GWp).

For comparison, this is about six times the 8 GWp of photovoltaic capacity already installed in France at the end of 2017. It is also more than twice the 20.6 GWp national target for 2023.

More and more projects for power plants on parking lots are developing, as shown by the calls for proposals from the Energy Regulation Commission on the realization and operation of electricity production facilities from solar energy: "Power plants on buildings, greenhouses and agricultural sheds and parking shades of power between 100 kWp and 8 MWp."

Figure 1.11. *The airport photovoltaic area of Montpellier (source: C. Grison)*

In 2015, photovoltaic shades were installed on a large parking lot at Montpellier Méditerranée airport. This system constitutes the largest airport photovoltaic plant in France. The 1,800 covered parking spaces have led to the installation of 16,016 anti-glare glass modules held by the shades. The plant produces 6,200 MWh of electricity annually, equivalent to the annual electricity consumption of 5,000 people. The energy required to manufacture a photovoltaic panel is offset by its production in 1 to 2 years. After about 25 years, the panel will be collected, dismantled and fully recycled.

1.10.3. *Creating solar fields in arid areas*

Arid zones are defined as tropical and temperate zones with an aridity index (or indicator of the degree of dryness) greater than 0.6. They include sub-humid, semi-arid, arid and hyper-arid ecosystems, such as Mediterranean scrub, steppes, savannahs and deserts. The solar resource is present everywhere but unevenly distributed. The geographical coordinates of many arid zones are ideal. Solar irradiation decreases as we get closer to the poles (or further from the Equator). Thus, many arid zones are strongly subjected to intense solar irradiation (such as the Atacama Desert in Chile and the Sahara Desert in Africa).

It is therefore easy to understand why Chile is positioned as a photovoltaic paradise. With the Atacama Desert, the sunshine conditions are ideal for developing photovoltaic fields. The altitude, the purity of the atmosphere and the relative coolness of this arid zone allow it to produce more than one-third of the time. EDF (a French multinational electricity utility company) has installed its largest park there.

Figure 1.12. *Large EDF solar park in the Atacama Desert, Chile; 700 hectare solar power plant (© Adobe Stock)*

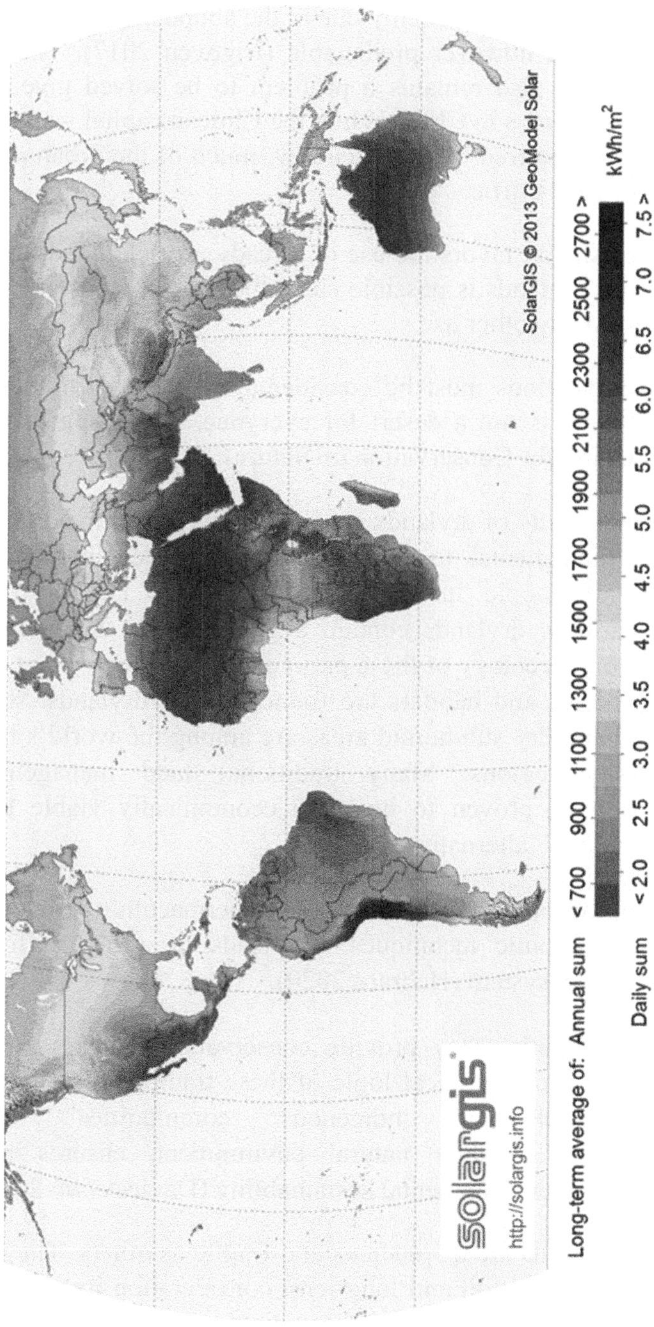

Figure 1.13. *World map of global solar irradiance (annual and daily average) (source: SolarGIS GeoModel Solar). For a color version of this figure, see www.iste.co.uk/grison/ecology.zip*

Initial feedback is not very encouraging. The performance obtained is far from the initial objectives, especially due to the abundant and sandy fine dust deposits, which were however predictable (Irigoyen 2017). The short life span of solar panels also remains a problem to be solved given the high concentration of UV rays in Chile. While the Chilean capital wants to have a metro that is 60% powered by green energy, much of the urban pollution is caused by heavy road traffic.

Although legislation favors the use of already artificial land, the creation of solar fields on drylands is possible since these open lands are unsuitable for agriculture and any other use.

Additional regulations must be considered for the use of these natural areas. An arid zone is not a desert for everyone. According to the IUCN (International Union for Conservation of Nature):

> [The] biodiversity of drylands is of enormous global importance, as it is fundamental to the well-being and development of millions of people in developing countries. Despite their relative aridity, drylands contain a great deal of biodiversity adapted to the ecology of these particular areas. Many plant and animal species and habitats are found only in drylands: some semi-arid and dry sub-humid areas are among the world's high biodiversity regions. Many traditional land management practices have proven to be more economically viable than many "modern" alternative practices.
>
> Thus, agroecology combined with permaculture involves ancient agronomic techniques and leads to a highly food-productive ecosystem (Hebrard 2020).
>
> More importantly, they provide conservation benefits at the same time. The ecological logic of these traditional strategies, developed through indigenous communities' deep understanding of their natural environment, ensures both economic and environmental sustainability (Davies *et al.* 2012).

These environments are as precious and fragile as others. They therefore deserve the same protection and long-term consideration in terms of nature conservation, species diversity and the maintenance of pastoral traditions.

The regulatory measures taken with regard to the end of life of installations and recycling of materials are certainly less restrictive than in Europe, but they are just as essential.

At a time when these ecosystems in arid zones are under unprecedented pressure, with no real solution to stop or slow down the degradation of ecosystems, it is appropriate to ask whether the installation of vast solar zones, at a distance from the beneficiaries, is really an opportunity. Pan-African projects such as the Great Green Wall, whose objective is to fight against the advance of the desert, are certainly more adapted because they are able to reconcile ecological regeneration, sustainable development and respect for the values and aspirations of local populations (Boetsch and Späni 2013).

1.10.4. *Creating photovoltaic parks on water*

On small island territories with a high urban density, the problem of energy independence is a real challenge. Innovative experiments have been carried out in this sense in Japan. Rich in water reserves (ponds and maritime areas), Japan is studying two systems of photovoltaic parks on water. Thanks to the ingenuity of French entrepreneurs from Ciel et Terre, a floating solar power plant has been created with the aim of contributing to the replacement of nuclear power plants. This plant of 4,500 solar panels covers a part of the artificial lake of Okegawa, in the northwest of Tokyo. The panels are based on floating modules made of high-density polyethylene (HD-PE), called Hydrelio®. This light, strong, inexpensive, flexible material is easy to assemble and can adapt to the undulatory movement of the waves. An optimal angle of inclination and surface covered seems to have been found to deploy a profitable system capable of supplying 400 homes. The coolness provided by the water (ideally 25°C) and the runoff of rainwater providing a natural washing of the panels are also performance arguments. This technology is now being deployed in at least 12 countries: Japan, the United States, the United Kingdom, China, Cambodia, Taiwan, Malaysia, Thailand, India, France, South Korea and Brazil (Ciel et Terre n.d.).

Further south in Japan, in Kagoshima, a different experiment has been conducted on a large scale. A power plant with 290,000 solar panels has been installed on an artificial island. Requiring a large flat surface, protected from typhoons by large concrete frames, this more conventional system does not have the temperature advantages of the floating system mentioned above. Finally, located near active volcanoes, the artificial island is regularly exposed to volcanic dust, which alters the capture of solar radiation.

Figure 1.14. *Omega 1 floating solar power plant in the commune of Piolenc (PACA region, France) (source: Akuo)*

These solutions complement island achievements such as the hydrowind power plant on the island of El Hierro, Spain, proving that energy independence is possible through the combination of renewable energies.

This technology is being developed in a disused quarry in Provence, where 17 ha of water are covered with 47,000 floating panels (L'Info durable 2019).

2

Photovoltaic Energy Production and Agricultural Activity: Agrivoltaics

Agrivoltaics aims to reconcile agricultural activity and energy production on the same surface. It is also known as agrisolar, photovoltaics of agricultural land or agro-photovoltaics.

What is at stake? Is it an opportunity to accompany the agricultural transition or an opportunistic solution that wants to monopolize agricultural land? Is it a useful protection against climatic hazards and certain severe episodes? Drought? Heat wave? Hail? Frost? Is it an opportunity to reduce the environmental footprint of agricultural systems or a threat to agricultural spaces? Is it compatible with the emergence of agro-ecology? Permaculture? Sustainable, eco-responsible and economically profitable agriculture? What are the risks and benefits?

Agriculture itself is changing. These important questions are at the heart of many debates and deserve to be asked.

The European Life AgriAdapt program (2016–2020) has characterized the vulnerability of European farms (field crops, livestock, perennial crops) and their adaptation to current changes (Meteyer 2020).

Between strategies capable of supporting the sustainability of farms and responding to climate change, demographic pressure and the reduction of arable land, the search for new solutions for adapting to a changing world is legitimate.

In order to provide food for thought, we propose to approach the agrivoltaic option through a few emblematic projects after having recalled the regulatory aspects.

2.1. Definition–legislation–adaptation and evolution of techniques

Article L111-1-2 of the French Urban Planning Code stipulates that photovoltaic installations incompatible with agricultural use are prohibited on agricultural land. This applies to all installations, including those that are not selected by a tender process.

Some photovoltaic projects are compatible with certain agricultural activities (e.g. innovative photovoltaic technologies whose development does not primarily affect the photosynthesis of crops). To assess this compatibility, the productive value of agricultural soils should be taken into account when considering the choice of location and the estimated production of the project. In this regard, the specifications of the MTES (*ministère de la Transition écologique et solidaire*, French Ministry of Ecological Transition and Solidarity) call for tenders on innovative solar installations that require the submission of a "technical brief on the synergy with agricultural use".

In order to respect this regulatory framework, techniques are constantly evolving.

Certain crops are to be favored when using these panels for an agrivoltaic project since some plant species have the ability to adapt to different conditions, such as growing in the shade. Lettuce has this ability, increasing the surface area of its leaves to perform more photosynthesis, which results in an increase in overall yield. Wheat, on the other hand, does not have this ability and would have a lower yield below solar panels.

A number of crops can be grown in the shade, but they would require a rigorous scientific study that takes into account all the advantages and disadvantages. In this regard, we would like to mention the Sun'Agri program led by the company Sun'R and INRAE (UMR G-eau, Montpellier) (INRA Écosolution Énergie 2017). This is a study planned over 13 years of research with specific objectives:

– *Sun'Agri 1 (2009–2013)*: studying the behavior of plants under fixed photovoltaic shading in various configurations and with a wide variety of crops;

– *Sun'Agri 2 (2014–2017)*: designing dynamic agrivoltaic systems;

– *Sun'Agri 3 (2018–2022)*: developing dynamic agrivoltaic systems on a commercial scale; developing for six representative species, optimizing the robustness of the control protocols and structuring of scientific research around dynamic agrivoltaic systems.

This study is interesting because the main theme is the design of new installations that will be useful for agronomy and that will enable Mediterranean crops to cope with climate change. It is therefore a question of proposing equipment adapted to the plants and not of focusing on energy and financial performance. Thus, techniques are constantly evolving. We present here a quick overview of the situation.

Agrivoltaic techniques vary according to the type and density of the panels, their inclination and their mobility.

Static, ground-mounted solar panels

There are two types of solar panel placement for a static panel agrivoltaic project. First, the panels can be at any distance from the ground, with agricultural crops between each row. The spacing between each row of panels must allow the various machines to pass through to easily carry out the different stages of agricultural cultivation.

Figure 2.1. *Photovoltaic farm with spaced static panels*

The second type of arrangement is to place the panels at a certain distance from the ground, in order to carry out the cultivation under these panels.

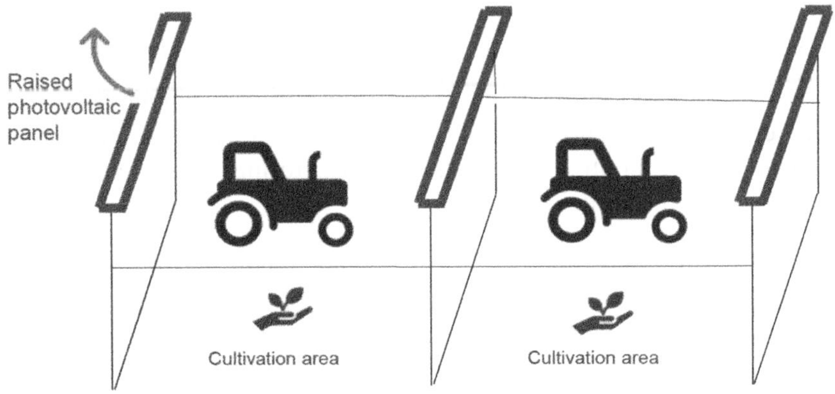

Figure 2.2. *Photovoltaic farm with modules mounted on piles with alternating cultivation areas*

When installing crops under solar panels, the reduction in radiation directly under the panels must be considered as well as for the surfaces around the panels (Marrou *et al.* 2013a). A non-exhaustive list of different techniques is presented below:

– *Agrinergie®*: plots of land between the panels of the solar fields are made available to farmers to interplant crops (called Agrinergie® first generation, Akuo Energy). Agronomic studies are necessary to determine the nature of the crops best suited to the device and the local environment;

– *vertical modules*: these allow cultivation between each row. The complementarity of solar and agricultural activities is ensured by a spacing of at least eight meters between each row of panels (Next2Sun process).

The panels can also be used as the fence of the property.

Due to its vertical structure, the technology is productive in the morning and afternoon and not at noon, while other more conventional installations operate at full capacity. On the environmental level, we can also question the effect of these installations on the landscape structure, air flows and the dispersal of pollinating insects.

Figure 2.3. *Example of vertical photovoltaic panels, solar farm in Donaueschingen-Aasen, Baden-Württemberg, Germany (source: Next2Sun). For a color version of this figure, see www.iste.co.uk/grison/ecology.zip*

Figure 2.4. *Solar park with vertical photovoltaic panels in Eppelborn, Dirmingen, Saarland, Germany (source: Next2Sun). For a color version of this figure, see www.iste.co.uk/grison/ecology.zip*

Dynamic solar panels, on the ground

These solar panels, called solar trackers, can be oriented or moved, in order to adapt the radiation received by them and by the crops, with the aim of optimizing the yields of both agricultural and energy productions. In order to optimize agricultural production, sensors measure in real time the sunshine received by the plants and the solar panels are oriented according to these conditions. Thus, it is possible to limit several climatic hazards such as excess or insufficient sunlight, water stress or spring frosts.

For example, the Sun'Agri experiment is based on this association of agricultural and energy production. The photovoltaic structure is built in such a way that all the maneuvers necessary for agricultural production can be carried out underneath the panels. As a result, the solar panels are placed 4 to 5 meters above the ground.

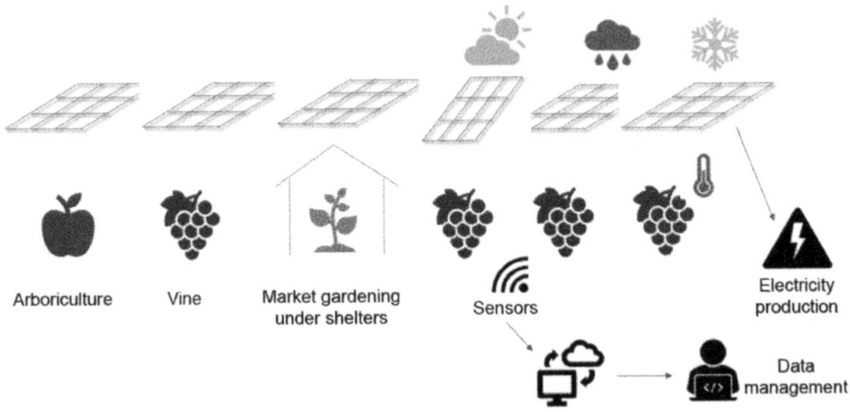

Figure 2.5. *Operation of the dynamic technology (example of Sun'Agri, France). For a color version of this figure, see www.iste.co.uk/grison/ecology.zip*

Some photographs of the field system are shown below (Figures 2.6A and 2.6B). These are corn crops grown for comparative purposes as part of the Sun'Agri program.

Dynamic photovoltaic shades are deployed to control the sunlight received by plants, in particular to limit sunburn, as well as to protect crops from heavy rain, hail and spring frost. On the other hand, they limit the natural supply of water by rainfall and can cause water stress if the crop is poorly chosen.

Cultivation in photovoltaic greenhouses

In France, in this type of installation, half of the greenhouse roof is covered with solar panels. This results in an overall reduction of 64% of the radiation received and, under the panels themselves, the reduction reaches 82%.

Figure 2.6A. *Different systems for combining agricultural production and electricity generation on the same plot (source: C. Grison)*

Figure 2.6B. *Different systems for combining agricultural production and electricity generation on the same plot (source: C. Grison)*

This significant reduction in sunlight has a direct impact on the growth of the plants to be installed and requires a judicious choice of the crops considered. This aspect will be detailed in section 2.2.

– *Greenhouses with semi-photovoltaic and anti-cyclonic coverings*: the last system that we present superimposes agricultural and solar production thanks to the development of greenhouses with semi-photovoltaic and anti-cyclonic coverings (installed in geographical areas suffering from violent cyclones). New Caledonia, which is particularly exposed to cyclones, benefits from the Focola project. It is based on the collaboration between a local farmer (J.C. Niautou), Akuo Energy and Enercal, which led to the construction of 16 anti-cyclone greenhouses, comprising 22,000 m^2, allowing permaculture market gardening activities. The power of the installation is 2 GW per year, which corresponds to the power supply of 600 homes.

This last advantage is a real asset for the French Overseas Departments and Territories (DOM-TOM), because it allows for progress towards energy independence and a more acceptable energy transition.

Figure 2.7. *First Agrienergy greenhouses in the Pacific, Focola site in New Caledonia, France (source: Akuo)*

2.2. Food crops

In this book, we present different experiences of shared activities through a varied panorama of agricultural activities: fish farming, oyster farming, viticulture, arboriculture, field crops, animal husbandry and market gardening.

For a more exhaustive study, it is advisable to consult the study carried out jointly by the companies Acthuel and Artifex (Acthuel 2020). It is an inventory of recent and unpublished data, existing projects, feedback from publications, press releases, reports and open-access websites.

2.2.1. *Fish farming*

Three fish farming projects have been identified in the French journal *Les techniques de l'ingénieur* (Combe 2020). We can mention, as examples, a sturgeon farm in Dordogne and a solar farm housing several fishponds on the island of Reunion (L'Écho du Solaire 2018).

The sturgeon farm, which is certified AquaRea (responsible aquaculture), includes 26,000 m^2 of ponds covered by a photovoltaic roof. The installation has a capacity of 5.06 MWp, which covers the electrical needs (100 kW) of the farm. The municipality of Neuvic-sur-l'Isle is the beneficiary of the surplus electricity production. Moreover, the installation attenuates the light stress undergone by the sturgeons. These conditions are closer to their natural living conditions (more than five meters deep) and therefore have little light. The shade of the panels also reduces the heating of the water by solar radiation, as well as the production of undesirable algae favored by solar radiation. The ultimate objective of this installation is the production of caviar (Verhaeghe and Blais n.d.).

On Reunion Island, the small habitable and buildable area is conducive to the optimization of the land and the accumulation of activities in a small space. Photovoltaic projects currently provide only 8% of the total electricity production, but the situation is changing rapidly. The Étang-Salé aquaculture farm corresponds to one and a half hectares of solar roofing, that is, 12 fish ponds, which protect the fish from the aggression of solar radiation and regulate the water temperature. Other advantages are described, such as the protection of fish from predatory birds, water savings and, especially, a 30% reduction in fish mortality.

In addition, a huge nine-hectare shed with solar panels is operated by EDF (Electricité de France). A complete study on the possible impacts on the environment (e.g. reduction of water filtration rate, landscape impact, etc.) and the benefits provided is available online (Biotope, Indian Ocean Agency 2013).

Figure 2.8. *Fishponds of the Etang-Salé solar aquaculture farm in Reunion Island, France (source: Akuo)*

In Asia, many similar projects are being studied. For example, the Vietnamese project Shrimps aims to cover land-based aquaculture systems with roofs made of solar panels. This type of photovoltaic fish farming project enables better management of water consumption by reducing evaporation, limiting inputs, protecting employees from the sun and protecting aquatic animals (fish, shellfish) from their natural predators.

2.2.2. *Oyster farming*

Various original projects have been developed. Among them is a project located in the Thau basin, near Montpellier (the project of F. Tarbouriech). Solar panels, associated with a battery, recreate the movement of the tides so that *"the oyster is forced to use its adductor muscle to remain hermetic and not to lose its water"*[1]. This mechanism should produce a fleshier oyster.

More recently, a vast project of floating solar shades has been imagined in the shellfish farming areas of the Thau basin, in order to solve a recurring problem: the *malaigue*, or proliferation of algae linked to heat waves, which causes a drop in oxygen levels in the water.

1 Quote from R. Tarbouriech, taken from the France TV info website: https://www.francetvinfo.fr/france/dans-la-lagune-de-thau-un-ostreiculteur-recree-des-marees_947319.html.

Figure 2.9. *Bassin de Thau, Occitanie, France: a place of innovation in the field of photovoltaics and shellfish farming activities (source: C. Grison)*

Again, the direct expectations are a decrease in solar aggression and surface water temperature.

Many additional innovative solutions are under consideration:

– solar energy supply of oxygen turbines to reinforce the oxygenation of the water;

– integration of fuel cells, which, by electrolysis of water, would produce hydrogen intended to supply the oyster boats.

All these solutions are being studied in other departments and could also be deployed in the DOM–TOM regions.

2.2.3. *Viticulture*

It is well-known that the grapevine needs sun, but too much sun produces wines with too much alcohol. A compromise can be found by using steerable photovoltaic panels. A first viticulture experiment has been carried out in the Pyrénées-Orientales (Domaine de Nidolères in Tresserre) (Hamdaoui 2019).

Similarly, in the Hérault region, INRAE has installed a similar system in the Florensac cooperative cellar, where sensors allow the panels to be oriented according to the climatic conditions and the needs of the vines.

While it is still too early to draw conclusions, the experiment deserves to be followed up and expanded.

Figure 2.10. *Solar panels installed in vineyards at Domaine de Nidolères, France (source: Sun'Agri)*

Another example can be found in the Vaucluse region, where for the past year, a 540 m^2 plot of land in the experimental vineyard of Piolenc has been protected from high temperatures by a solar roof. Here again, the position of the panels is adjusted according to the amount of sunshine: the plants benefit from the sun in the morning and are protected at the hottest times of the afternoon. The first results of this agrivoltaic experiment show a reduction in water stress, allowing for a prolonged growth of the vines and the formation of heavier berries (17% increase in mass) (Vitisphere 2020). This result is logical, as it is known that lack of water impacts photosynthesis, shoot growth and berry size. The anthocyanin content is higher (+13%), as is the total acidity (+9% to +14% depending on the modality).

The increase in temperature induces a higher sugar concentration in the grape but a lower concentration of secondary metabolites such as natural pigments, affecting its color. Thus, the shade provided helps to preserve the aromatic profiles of the wines produced. The protection of the vine crops increases the quality of the resulting grape (Van Leeuwen and Darriet 2016).

Early harvesting is also avoided.

On the other hand, the decrease in overall sunlight results in a lower concentration of carbohydrates in the grapes, which results in a decrease in the alcohol content of the wines obtained.

It will be interesting to know the impact (positive or negative) of the lack of sunlight on the development of parasitic fungi of the vine and on the functioning of the plant's natural defense systems.

2.2.4. *Arboriculture and fruit production*

The combination of arboriculture and photovoltaics requires specific arrangements:

– tree trellising thanks to the structure's posts (one tree row out of two);

– installation of protective nets in the orchard (hail, insects, wind);

– automatic net unwinding and winding systems coupled with the presence of an on-site weather forecasting station.

Experiments in fruit growing (citrus, avocado, manioc, etc.) have been reported in Reunion Island (Ligne des 400) and in New Caledonia. In metropolitan France, agrivoltaic systems in various fruit crops (apricots, apples and pears) are already operational. A pilot experiment has been carried out on a former 9 ha orchard in Bellegarde (Gard), a partnership between Akuo Energy and MAIF. The site has been rehabilitated and has benefited from photovoltaic shading.

The provision of suitable shade is an alternative to irrigation in dry periods: 20% water savings are possible on a plot. The protection of the fruit against burning during periods of strong sunlight is the major advantage, particularly in the south of France. In the case of apricot production, another important advantage has been highlighted: the absence of rainwater limits the development of cryptogamic diseases such as Monilinia and oidium. It is thus possible to avoid phytosanitary treatments based on sulfur and copper.

For the experimenters, the reduction of erratic losses due to climatic hazards (frost, hail, strong wind) compensates for a lower production (-30% of fruits on average) due to the lack of light.

On the other hand, the nutritional composition of the fruits is partially affected by the growth of the trees under shade; therefore, under solar panels, the carbohydrate concentrations are lower, as well as the levels of N, Mg, S, Mn and Na (Hirzel *et al.* 2020).

Very recently, agrivoltaic projects have started for fruit production with innovative semi-transparent modules (like those of the company BayWa r.e.). These semi-transparent solar modules are designed to compensate the strong reduction of solar radiation for the plants while protecting them from heavy rain and direct light. The first experiment with raspberry production seems to be successful and is leading to other similar projects in the Netherlands for the cultivation of berries such as red currants, blueberries, lingonberries, blackberries and strawberries.

Similar experiments are starting in France, such as the La Pugère experimental station (Mallemort, Bouches-du-Rhône) for apple and pear trees. Projects are underway to extend this experience to an almond orchard (Reconception of the almond orchard for a successful transition to agroecology (Revatra)).

Agrivoltaics also seems to be well-adapted to kiwifruit production. As an example, we can mention the device initiated by the Chambers of Agriculture within the framework of a collaboration between the agricultural high school of Sainte-Livrade-sur-Lot (Lot-et-Garonne) and the company Reden Solar (Innov'Action project). The high school has developed a hectare of photovoltaic greenhouses with a total power of 1 MWp to study new growing conditions. The greenhouses are equipped with mechanical openings on the roof and façade. Strips of light and shadow sweep across the ground surface of a greenhouse to homogenize the light and thus the maturity of the fruit.

Figure 2.11. *Strawberry cultivation under photovoltaic panels; Babberich, Holland (source: BayWa-Agri)*

One last example is in the Isère region, in Jarcieu, where the company Reden Solar has extended its experiment to a total solar surface area of 9,515 m^2. For an equivalent area, agricultural production would be at least equivalent to that of an open-field crop, or even higher after solving the classic problems of light and ventilation encountered in greenhouses. The crops of strawberries, potatoes, lamb's lettuce and raspberries are favored.

The conclusions of these first experiments in agrivoltaics applied to fruit crops are motivating and indicate that they could be adapted to organic cultivation in photovoltaic greenhouses, such as for tomatoes, cucumbers, eggplants, peppers, potatoes, lamb's lettuce, zucchini, asparagus, raspberries, open-ground or suspended strawberries, kiwis and aromatic plants (Polese 2006).

If these pilot experiments prove the need to find conditions adapted to these photovoltaic greenhouses, the results are encouraging enough to extend the concept to other crops. For example, Germany, which is a very dynamic country in this field, has already tested the cultivation of winter wheat, potatoes, celery and clover as part of the Agrophotovoltaik project (L'Echo du Solaire 2017).

2.2.5. *Field crops*

Projects in field crops become possible with the development of light and resistant aerial structures.

An agrivoltaic solar tracking system (Peace and Christopher 2019) was examined in combination with a corn crop in a simulation study. For this purpose, a software platform was developed coupling a radiation and shading model to the GECROS generic crop growth simulator. The simulation was conducted using a 40 year climate dataset from a corn-growing location in northern Italy. Different parameters were studied (panel density, sun intensity, humidity, etc.). Simulations for an irrigated corn crop in full light, under conventional conditions, have been carried out for comparative purposes. The main conclusions are that the reduction of solar radiation by the photovoltaic shading:

- significantly reduces the average temperature of the ground;
- significantly reduces evapotranspiration;
- can maintain a water balance in the soil.

From this perspective, conditions are more favorable for plant growth than in full sunlight, and average grain yields are higher and more consistent under photovoltaic panels than in full sunlight. The advantage of growing corn in the shade is enhanced during periods of drought, indicating that agrivoltaic systems may increase crop resilience to climate change.

In Germany, field experiments have shown the possibility of growing potatoes under photovoltaic panels. Hot summers enhance the comparative performance of conventional and solar field crops.

These results are supported by a U.S. study (Barron-Gafford *et al.* 2019) on bell pepper and cherry tomato crops, which provides evidence that the shade provided by the photovoltaic panels decreases the drought stress of the plants and reduces the heat of the photovoltaic system. Agricultural production and electricity can therefore be synergistic. The results presented provide further food for thought and motivation for future explorations towards resilience of food and energy systems under increased environmental stress involving heat and drought.

While potato crops do not require much sunlight, the situation is different for wheat and corn. Experiments in Japan have solved this problem with spaced panel installations.

Elevated systems, or those with bifacial solar modules, that is, capable of capturing the direct solar radiation arriving on the front face as well as the radiation reflected by the back face, will undoubtedly allow the cultivation of more complex species while respecting their photosynthetic activity.

Recent ambitious studies have been conducted in China. Various applications of photovoltaic agriculture have been combined with photovoltaic greenhouses, photovoltaic animal husbandry, wastewater purification, photovoltaic water pumping and rural solar power plants. Nevertheless, more theoretical research should be conducted alongside experiments to optimize the combination of photovoltaic power generation and farming in relation to the environment (Xue 2017).

However, a study conducted in China on the economic and social performance of five large-scale agricultural greenhouses expresses a number of reservations about the economic and social impacts of such installations. The feed-in tariff is much lower than the model predicts. This implies that photovoltaic farms should pay more attention to crop planting and product quality and that policymakers should focus their financial incentives more on the quality and yield of agricultural production, not only on photovoltaic power generation (Li *et al.* 2017).

2.2.6. *Market gardening: shades, greenhouses, protective structures for environmental conditions*

If there is a conflict of use between the agricultural world and the photovoltaic power plants on the ground, the situation seems to be different for greenhouses, in particular for market gardening:

– the use of heated greenhouses is subject to the increase in the cost of energy; market gardening in a cold greenhouse in a more favorable ecological framework can be an opportunity;

– the economic conditions of most farms no longer allow for significant investments to experiment with new technologies.

The development of greenhouses with secure financing through the resale of electricity could perhaps revive this sector by targeting new objectives:

– developing a quality and seasonal production, with a low carbon footprint;

– protecting market garden productions (strawberries and other red fruits, tomatoes, carrots, etc.) from climatic hazards (bad weather, late frosts or too much sun), from pests and from cryptogamic diseases by reducing humidity, and thus reducing phytosanitary products;

– improving the working conditions of market gardeners;

– supporting local channels and small farms in organic reconversion;

– managing runoff and irrigation/spraying systems in a more optimal way.

These benefits can be achieved with highly technological greenhouses that are redesigned to meet these objectives. Advances in the field include systems that extend and retract to modulate shading and adjust climatic parameters such as humidity, temperature and light.

Advanced research has been carried out to characterize and model the impact of the tilt of dynamic solar panels on the water balance of a lettuce crop using a dynamic agrivoltaic system. Reference can be made to the thesis of Y. Elamri (2017) for a comprehensive study on the water balance of these installations or to the work of B. Valle (2017) on the potential of steerable panels for maximizing the productivity of lettuce crops and the phenotypic plasticity of lettuce in the shade of photovoltaic panels.

2.2.7. *Other crops*

2.2.7.1. *Crops for non-food use*

2.2.7.1.1. Production of essential oils

Since 2010, agrivoltaic projects have been underway in Reunion Island. On a territory where the spaces for cultivation are limited, the technique is well-adapted. In order to optimize the cultivation space, the rows of panels have been deliberately spaced to allow the cultivation of intermediate rows. Thus, a crop of lemongrass, traditional on Reunion Island, complements the crops under panels (Agrinergie® project 2010).

Figure 2.12. *Solar field and lemongrass cultivation in Pierrefonds, Reunion Island, France (source: Akuo Energy)*

2.2.7.1.2. Horticulture

Extreme weather and pest pressure impact the viability of horticultural operations. These crops, already subject to strong seasonal fluctuations, require comprehensive and efficient protection. To compensate for rapid weather changes, a solar roof system opens and closes to reproduce the right conditions for plants to thrive (Crop Climate Control 2020). Flowers are protected and perennial plants are more resilient to weather events. In winter, soil temperatures can drop by as much as 30% below the annual average. Dormancy breaking and flower quality are facilitated.

Educational experiments have been initiated. A horticultural project was carried out jointly with the Compagnie Nationale du Rhône and an agricultural training establishment specializing in horticulture. The aim of this project was to demonstrate that the microclimate created by the photovoltaic structure above the cultivation areas protects them, increasing productivity and decreasing water consumption (Boulanger 2020).

2.2.7.2. *Mushroom cultures*

The shaded, humid and warm areas under the photovoltaic panels seem to offer favorable conditions for the cultivation of mushrooms. In the experiment conducted by the company Sustainergy, sensors constantly monitor the degree of humidity. There are plans to expand this project to two farms that will produce each year (Sustainable Decisions 2017):

– 40 tons of edible mushrooms (Judas ears);

– 4,000 kWh of solar energy, sold to a local supplier.

2.2.8. *Pasture and livestock*

Combining animal husbandry with the maintenance of the surfaces on which the solar panels are installed is an idea that is benefiting from the beginning of feedback. It concerns mainly sheep and more rarely poultry. Goat and pig farms are not very developed because of the risks of degradation of the installations (e.g. goats jump easily on the panels). A major aspect concerns the maintenance of the photovoltaic park plots by sheep. Two points of vigilance are to be noted:

– sheep should not be used to reduce the cost of soil maintenance alone;

– solar photovoltaic fields should not be built near substations of interest (transformation centers, management and distribution of electricity), on fertile agricultural land or near access to water resources.

A study was conducted on the environmental effects of solar panels on a non-irrigated pasture that is often water-stressed. Changes in microclimatology, soil moisture, water use and biomass productivity due to the presence of solar panels were quantified. This study demonstrated the importance of considering the impacts of these factors in the design of solar farms to take advantage of potential net gains in agricultural production.

Microclimate stations were placed underneath agrivoltaic solar panels in a campus in Oregon, USA, two years after solar panel installation. Soil moisture was quantified using probe readings. Significant effects of solar PV panels on average air temperature, relative humidity, wind speed, wind direction and soil moisture were observed. Areas under solar PV panels maintained higher soil moisture throughout the observation period. A significant increase in late-season plant biomass was also observed for the areas under the photovoltaic panels (90% more biomass). Finally, soils under the photovoltaic panels had significantly greater water capacity (+328%) (Hassanpour Adeh *et al.* 2018; Figure 2.14).

Figure 2.13. *Photovoltaics and livestock; maintenance of a park by a flock of sheep (source: DAVELE)*

The co-activity must be based on the progressive construction of these projects with the expertise of agricultural engineers over the long term.

A ground-mounted photovoltaic park can go hand in hand with the maintenance of a pastoral activity if the height of the structures and the layout of the various elements have anticipated this use. The photovoltaic developer can also include in their project the preparation of meadows (possible earthworks, seeding, etc.), the installation of water points for the animals (Hespul national photovoltaic resource center) and fodder supplements.

As an example, we would like to mention the experience in Ortaffa (Pyrénées-Orientales). The company Juwi EnR (now Neoen) and the mayor of the municipality have developed a specific approach corresponding to the reality in the field of a photovoltaic park of 25 MWp. The park, which was connected in September 2013, is located on 87 hectares of land of low agronomic value, such as wasteland and vineyards classified as VCC1 (excluding AOC (*appellation d'origine controlée*, controlled designation of origin)), for the major part of the project. The facilities have been planned according to the nature of the livestock. Large areas of grass are available for them.

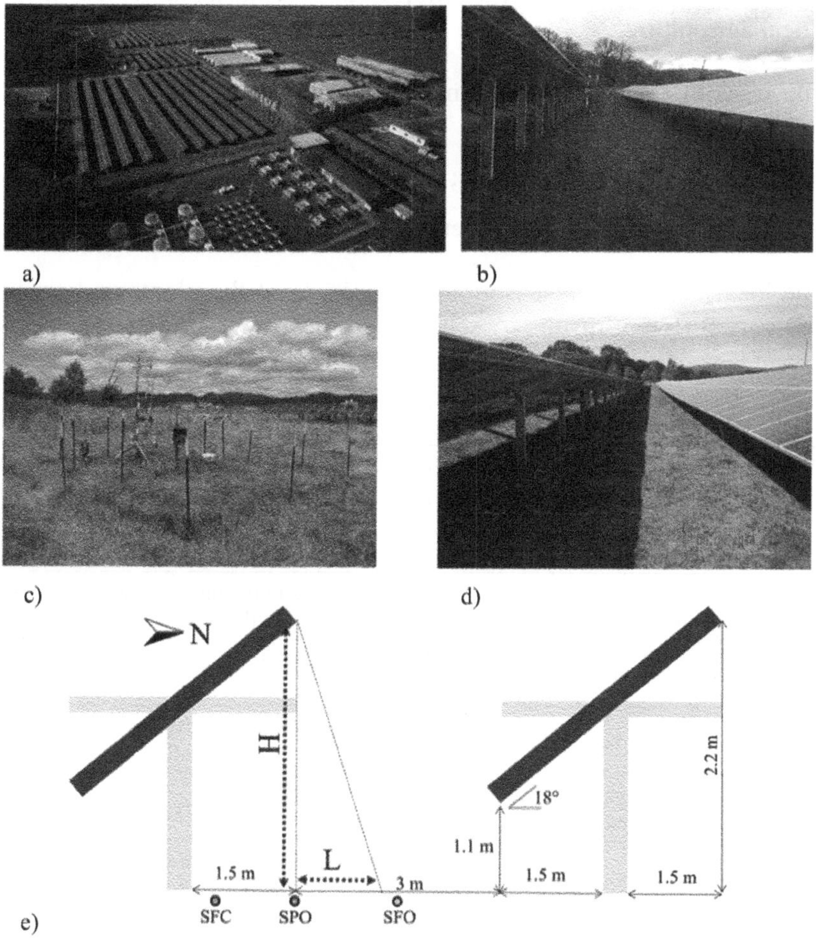

Figure 2.14. *Influence of a photovoltaic system on soil moisture and water use efficiency (Hassanpour Adeh* et al. *2018)*

The development of an outdoor poultry activity (chickens, geese) is interesting. It makes it possible to meet the CAP (Common Agriculture Policy) specifications for an AB (organic) label with 30% shade provided by the panels by adopting an agroforestry model. Little developed in France, it is more developed in Japan, the United Kingdom and the United States (Now 2019).

In conclusion, photovoltaic technologies can be used in a wide variety of agricultural applications. Numerous advantages are identified, especially in times of global warming and transition to a more sustainable, less intensive and local agriculture.

The transition from conventional agriculture to agrivoltaics means two major changes:

– reduced sunlight on the crops;

– reduced contact with rainwater.

The local climate, the season and the nature of the plants and crops can lead to very different results. A schematic breakdown of the advantages and disadvantages of agrivoltaic crops is given in Table 2.1.

Advantages	Disadvantages
Protective shade in hot weather	Decreased yields due to lack of sunlight. The systems with inclination and orientation limit this difficulty. Staggered harvest dates
Protection during severe weather and violent climatic events	Heterogeneity of sunshine depending on the place of the crop under the plot
Limitation of evapotranspiration	Modification of the nutritional composition of food products
Protection against hail and frost	Difficulty of mechanization (mitigated by the high installations)
Reduction of cryptogamic or fungal diseases	Reduction of the cultivable space
Reduction of the amount of watering	Heterogeneous distribution of rainwater at the soil level (erosion)
Protection against certain volatile predators	Decreased pollination

Table 2.1. *Schematic balance between the advantages and disadvantages of agrivoltaic crops*

This assessment should be qualified according to the plant species used. Lettuce, spinach, parsley, broccoli and beans do not seem to be affected by the lack of sunlight. The problem is more delicate with melon, tomato or cucumber; this problem is easily solved in permaculture, which makes it possible to exploit this difference by placing the various plants in the right place.

In recent installations, a compromise is sought between maximum energy production and an installation that allows part of the solar radiation to pass through for the growth of plants. It is defined by the land equivalent ratio (LER) index:

$$\mathrm{LER} = \frac{\text{Harvest yield in AV}}{\text{Harvest yield in MC}} + \frac{\text{Electricity yield in AV}}{\text{Electricity yield in PV}}$$

where AV stands for "agrivoltaic system", PV stands for "conventional photovoltaic system" and MC stands for "conventional monoculture".

The LER must be greater than 1; agricultural production and electricity systems are higher than in separate systems (Dupraz *et al.* 2011; Marrou *et al.* 2013a, 2013b; Fu *et al.* 2019; Sekiyama and Nagashima 2019).

Beyond the balance to be found between sunlight for crops and performance in electricity production, the profitability of ground-mounted photovoltaic power plants is not obvious. It is conditioned by a large surface area (from 1 to 2 ha for 1.25 MW according to ADEME), which may seem contradictory with modern agriculture and may lead to conflicts of use with agricultural or forest land. In order to prevent these conflicts, the choice of location should be made in priority on non-vegetated surfaces unsuitable for agriculture (industrial wastelands, former quarries, degraded and not very fertile sites, industrial zones, etc.). The projects of photovoltaic power plants can integrate a mix of agricultural activities, such as agrivoltaics. Thus, some animal production sites (fish farming, oyster farming, extensive sheep or goat husbandry) and plant production sites (market gardening, etc.) are compatible with ground-mounted photovoltaic power plants with suitable structures. Thorough research into the best growing conditions and the smallest possible environmental footprint is necessary before any agrivoltaic project.

3

Innovative Principle of Ecovoltaics

3.1. Definition and concept

The general objective of ecovoltaics is to valorize the surfaces currently not exploited under the solar panels of photovoltaic parks in a scientific ecology program with an applied aim. From a global point of view, the aim is to find coherent solutions between climate change control, creation of conditions for a responsible and sustainable use of resources and energy, and preservation of biodiversity.

If it is accepted that agrivoltaics can be an opportunity for agricultural transition, it seems interesting to show that photovoltaics can also be an opportunity to develop unusual ecological solutions. The approaches presented here were initiated as part of a unique collaboration between the *Société monégasque de l'électricité et du gaz* (SMEG) and the laboratory of bio-inspired chemistry and ecological innovations (ChimEco) of the CNRS and the University of Montpellier for the Boissière photovoltaic park (Gard).

The objective of this project was to promote the installation of plants under photovoltaic panels. The following types of plants were selected for their ecological properties:

– plants for biodynamic agriculture;

– melliferous plants, attracting bees and other pollinators;

– plant species threatened with extinction (classified in the red list by the IUCN);

– plants of medicinal interest.

3.2. State of the art: feedback

Ecovoltaics is a completely new concept inspired by permaculture. The objective of this chapter is to gather the available information on this new way of using the soil under photovoltaic equipment in order to provide initial feedback that will serve as a basis for new projects.

In order to conceive a new valorization of useless solar spaces, knowledge of the installed plant species and their future use is essential. In addition to the field experiments, a bibliographic overview of plant species and their properties is presented.

The conditions of cultivation correspond to a difficult case: poor and not very fertile ground, fixed and non-orientable solar panels, panels with low heights (from 1 to 3 m) and a site subjected to the droughts of the Gard and exposed to wind. This first experiment was therefore a challenge for cultures under solar panels.

Figure 3.1. *Experimental site under solar panels in Boissière (Gard, Occitanie) (source: C. Grison).*

3.2.1. *Constraints of the experimental site in Occitanie*

The constraint of the soil, the absence of direct light, the lack of water and the Mediterranean heat led to the selection of plants adapted to these cultivation restraints.

The plants chosen for the experimentation had to meet different characteristics:

– these plant species had to not have a high light requirement, since they would be in the shadow of the non-dynamic photovoltaic structures;

– they had to also develop a modest herbaceous or, for some, bushy habit, because the panels were placed at 1 m from the ground on the lowest side (south) and 3 m on the highest side (north);

– they had to also fertilize the soil as a double objective:

– the plants had to be installed in a perennial way without tilling the soil;

– the plants had to enrich the soil with nutrients in a sustainable way from the perspective of the end of the photovoltaic activity.

Following the book *Je cultive en lasagnes partout et toute l'année* (Lapouge-Déjean *et al.* 2018), the substrate for the plants under the solar panels was constructed using the so-called "lasagna" technique. This permaculture concept, invented in the 1980s by Patricia Lanza, involves placing successive layers of different substrates to replicate humus-rich soil, mostly using reclaimed plant waste. The goal is to promote the natural decomposition of the different elements to mimic the humic soils of forests. Once installed, this substrate provides plants with the necessary nutrients, as well as promoting water conservation and easy root development. The advantage of this principle is that it can be used anywhere, including on non-fertile soils or concrete, and does not require maintenance.

The basic principle is to alternate layers of so-called "green" materials (freshly cut plants, grass clippings, compost, fresh animal excrement, etc.) which are rich in nitrogen and water and serve to initiate the mineralization of the elements (driving effect) with so-called "brown" materials (dry leaves, wood shavings, straw, cardboard, etc.), which serve as a substrate for the mineralization in the long term and are rich in carbon.

These layers were simply placed on the ground (without any major work beforehand) and are 30 to 50 cm high. The green materials in these layers were very rich in nitrogen. Around these layers, on the ground, were placed stones to prevent all the elements of the lasagna falling down by sliding on the edges.

There are many different types of lasagnas, depending on the type of planting sought, the type of soil on which they are planted and the expectations of the growers. For this experiment, the choice was made for dry and very poor soils in which plants would have had great difficulty in developing. The area chosen was 45 m^2. We placed the layers as shown in Table 3.1.

1) Cardboard	5 cm
2) Wood chips	5 cm
3) Horse manure	5 cm
4) Straw	5 cm
5) Fresh compost	5 cm
6) Wood chips	5 cm
7) Aged compost	5 cm
8) Potting soil (plus clay pebbles)	5 cm (plus two handles)

Table 3.1. *Structure of the lasagna installed under the solar panels in Boissière (Gard)*

Each part of this substrate has a specific purpose.

– *Cardboard*: the cardboard is on the ground and covers it. This way, it prevents weed seeds present in the soil from having access to the rich substrate intended to feed the plantations. Moreover, the lignin present in the cardboard serves as a substrate for soil microorganisms.

– *Wood chips*: the wood chips' main role is to bring carbon to the lasagna. They are the element richest in carbon that we can put in a soil (C/N = 100 to 400). Their microbial decomposition is also a source of moisture. In addition, the relatively coarse elements allow for aeration of the substrate, since it is very important to prevent the roots from suffocating.

– *Horse manure*: fresh horse manure collected the day before planting acts as green matter. It is a material rich in water that also allows a very good contribution of nitrogen to the soil.

– *Straw*: straw serves as a carbon-rich element (C/N = 100) and has the same overall role as wood shavings, since its rigid structure allows it to retain a significant amount of air in the substrate and its carbon content allows it to serve as an important source of nutrients for the soil's microbial flora, but it will degrade more quickly.

– *Fresh compost*: the mineralization stage of this compost is not yet completed. Therefore, the nutrients are not yet available for absorption by the plants. Placing it at this level of the lasagna layers gives it time to begin the mineralization phase before the roots have access to it. This compost becomes useful when the plants have been able to establish themselves and the roots have already penetrated the first layers of substrate.

– *Aged compost*: this compost already has many minerals available to plants. Mineralization is not yet complete, but it will be sufficiently advanced when the plants develop their roots in this layer of compost to provide the plants with the energy they need to grow.

– *Soil*: the seedlings are directly planted in the soil. It is a soil that is very rich in immediately available organic matter so that the young plants can immediately draw the elements necessary for their survival. However, this layer of soil will be depleted of its available elements faster than the lower layers, since they will be drawn off as soon as they are planted. The clay pebbles provide mineral matter, which is essential for the quality of the soil. In particular, they allow the formation of clay–humus complexes with the organic matter present in quantity in the soil. This prevents all the elements present in the soil from being solubilized in the water and evacuated by runoff.

The concept of a substrate in different superimposed layers allows the soil to be supplied with all the elements normally brought to the crop through amendments, but in one go. The concept is also designed to limit the development of weeds (presence of cardboard on the ground). Finally, the main objective of lasagna cultivation is to bring significant microbial life to poor soils by imitating a natural environment in which it is very present (Lapouge-Déjean *et al.* 2018).

Figure 3.2. *Lasagna technique developed by B. Lapouge-Déjean (Lapouge-Déjean* et al. *2018) used under solar panels. For a color version of this figure, see www.iste.co.uk/grison/ecology.zip*

Figure 3.3. *Lasagna installation under solar panels (source: C. Grison)*

3.2.2. *Planting of species of interest in biodynamic agriculture on the grounds of solar parks*

There are very few scientific studies on the usefulness of plants of interest in biodynamic agriculture. The objective here is, in addition to enhancing the value of the solar parks, to make a reliable scientific contribution concerning the usefulness of certain natural substances in phytosanitary treatments considered traditional (slurry, decoctions, herbal teas, etc.).

In this research, five species were selected:

– eagle fern (*Pteridium aquilinum* (L.) Kuhn);

– alder buckthorn (*Frangula alnus* Mill.);

– comfrey (*Symphytum peregrinum* Ledeb.);

– garden rhubarb (*Rheum rhabarbarum* L.);

– borage (*Borrago officinalis* L.).

The objective of this work is not only to show that it is possible to cultivate plants useful for the preparations used in biodynamic agriculture under photovoltaic panels but also to bring elements of knowledge that will allow for progress in the use of these practices and to complete the know-how of the farmers and the passionate persons of the field with a more theoretical approach and a complementary perspective.

Hence, the botanical and ecological characteristics and the active molecules of these plants are presented successively, species by species.

3.2.2.1. *The eagle fern,* Pteridium aquilinum

3.2.2.1.1. Botanical characteristics

Pteridium aquilinum is a species of the Dennstaedtiaceae family. It grows mainly in the undergrowth, but it is also found in meadows, ditches, stream banks, etc. The eagle fern adapts very easily to many environments. It has, for example, an unusual tolerance for semi-shaded or even sunny areas. It is generally located on acidic soils, where many other plants do not survive (the optimum pH of this species is less than 4.5) but can grow on alkaline and carbon-rich soils. The total biomass of stands of *P. aquilinum* can reach up to 12 t/ha of dry matter.

What is the geographical distribution?

Because of its adaptability, this herbaceous plant is said to be cosmopolitan (Pelt 2011), meaning that it is geographically distributed virtually worldwide.

The eagle fern is found on every continent except Antarctica. It is the fifth most abundant plant in the world (Kisielius *et al.* 2020). In addition, a notable feature of *P. aquilinum* is its natural biological activity against pests, which allows the eagle fern to protect itself from most predators.

Development under solar panels: an opportunity?

Although its development is more extensive in the sun, it resists very well in the shade. According to Dumas (2002), this plant species does not need more than 2% of incident radiation to survive. On the other hand, it does not regulate its stomatal conductance as fast as trees, which makes it more vulnerable in cases of lack of water and affects its tolerance to drought.

Easily installed in spring under the solar panels, the local semi-shade suited it perfectly. Only regular drip watering was necessary. The lasagna effect was beneficial to it because it conserved humidity.

Figure 3.4. *Eagle fern growing under solar panels in Boissière (Gard) (source: C. Grison)*

3.2.2.1.2. Use in biodynamic agriculture

Eagle fern is mainly used in the form of mash (maceration). In this form, it has insecticidal properties against many pests, such as aphids, woolly aphids, leafhoppers and mealybugs, as well as wireworms and slugs (Masson 2018).

It is also used as a mulch of about 15 cm, providing good protection against frost. It is also said to act as a fungicide on gray mold (*Botrytis cinerea*) of strawberries. It is also in this form that it is the most effective against slugs. Indeed, the latter appreciates eating eagle fern leaves, despite their strong toxicity towards them.

According to the existing literature on biodynamic agriculture, the fronds should be collected before the opening of the sporangia, that is, before August. The maceration can be used pure or diluted, depending on what is to be treated.

Since eagle fern leaves are rich in nitrogen and potassium, they can also be used as a green manure, mulching the soil. This also keeps the soil moist and covered to prevent weeds from growing.

Finally, some also attribute fungicidal properties to these leaves when they are used in an infusion (Agrobio 2012).

On the other hand, the whole plant has toxic effects on animals. Indeed, several cases of death in herds of sheep or cattle have been recorded when food was lacking in the pastures and they were forced to eat eagle ferns. Different molecules are responsible for this toxicity, which is also effective against many insects (including phytophages) (Dumas 2002). One of the main interests of derivatives of this plant is that it is one of the few natural curative treatments. Given its possible toxicity, it is worth asking about the specific secondary metabolites that the eagle fern uses as defense molecules.

A detailed study of the secondary metabolites used as defense molecules is presented in Appendix 1. The mechanisms of action of prunasin, ptaquiloside (also called *aquilide A*) and thiaminase I (antithiamine) are developed. They reflect the mode of action of eagle fern in the form of slurry or mulch. The expected effectiveness of these molecules should also encourage caution and the rational use of these natural solutions. It would be interesting to study these data in depth and to propose protocols for their use, as is done conventionally for drugs. This solution would make it possible to reassure the authorities who issue marketing authorizations for this type of compound, as well as to advise future users.

3.2.2.2. *Alder buckthorn,* Frangula alnus

3.2.2.2.1. Botanical characteristics

Alder buckthorn is a shrub of the family Rhamnaceae that generally grows in very wet ground, even directly in rivers. Native to Eurasia, this species is quite resistant and has the advantage of also having an ecological interest, with flowers that attract many insects (and in particular bees). While it is very resistant to cold and humidity (it is recorded as an invasive species in North America where it has been imported), the alder buckthorn does not tolerate drought, which is a major constraint for planting it in the Mediterranean region (Reinartz and Kline 1988; Hamlin *et al.* 2015).

Figure 3.5. *Growing alder buckthorn under solar panels in Boissière (Gard) (source: C. Grison)*

It can withstand both light and shade, and its total biomass does not vary greatly depending on its exposure. On the other hand, its architecture varies a lot. It preferentially adopts a tree-like form when it is under an existing canopy, whereas in full sun, it tends to present a shrubby form. Its total biomass approaches 5,285 kg/ha in a plantation if the planes are spaced 3 m apart in length and 4 m in width (Hamelin *et al.* 2015).

Its development at the edge of the solar panels is ideal provided that the wind does not infiltrate under the panels.

3.2.2.2.2. Use in biodynamic agriculture

In addition to the fact that the alder buckthorn is a melliferous plant, its bark is used for its fungicidal properties, especially against mildew. For grapevine, it is used in the form of a decoction as an elicitor, in order to stimulate the natural defenses of the plant.

For the moment, the tests concern the decoction of alder buckthorn coupled with a very reduced dose of copper. Indeed, the classic treatment of mildew is usually done with copper alone, either in the form of Bordeaux mixture or in powder form. Here, the results of the Viticultural Technical Association (Bulletin technique viticulture 2013) show that a dose of 500 g/ha of copper does not have statistically more effective results than a dose of 150 g/ha of copper coupled with a dose of 150 g/ha of alder buckthorn extract.

Using alder buckthorn, the biosynthesis of the molecules of defense of grapevine is multiplied by 25 in cases of attack by the mildew. On the other hand, it is considered as a molecule with a priming effect, which means that it has no effect on the vine before the attack, but only after. It is therefore only a curative treatment (Gindro *et al.* 2007).

Mechanistic details are presented in Appendix 2 of this book for the reader who would like to deepen their knowledge of the mode of action of alder buckthorn in biodynamic agriculture.

3.2.2.3. *Comfrey,* Symphytum peregrinum

3.2.2.3.1. Botanical characteristics

Comfrey is a Boraginaceae commonly found in Europe and part of Asia. It has rapidly dispersed within North America where it has been introduced, as well as in South America. In France, its range is throughout the northern part of France, while it is quite rare in the Mediterranean region (Staiger 2012). This can be explained by its high need for water. This species is more likely to grow in moist soil and sunlight. Therefore, it is frequently found on the edge of streams, on the edge of forests or in the fresh and rich soils of meadows.

Of a herbaceous nature, it generally measures between 60 cm and 1 m in height. It has a large and voluminous rhizome and reproduces essentially vegetatively. Its bell-shaped flowers are grouped in inflorescences, which attract pollinators. Honeybees have little access to the pollen until the base of the bell-shaped corolla has been pierced by another insect (bumblebee, beetle, etc.).

Comfrey has been known since antiquity for its pharmaceutical properties, especially for its healing activity. It was often consumed in salads, young shoots and herbal tea, but the toxicity of the pyrrolizidine alkaloids present in the leaves and rhizomes made it banned from many pharmaceutical recipes. Indeed, the hepatotoxicity of these compounds in cases of prolonged consumption is known. On the other hand, they do not penetrate the skin; therefore, the traditional recipes concerning the cutaneous applications did not disappear (Évreux 1989).

3.2.2.3.2. Use in biodynamic agriculture

In addition to its pharmacological benefits, comfrey is also widely used by lovers of natural phytosanitary products. Indeed, while this plant is very rich in nitrogen, iron, silica and potassium, it is especially its high content of interesting trace elements that makes its reputation. It allows an interesting amendment with zinc, boron or manganese.

Like nettle, comfrey is often used in the form of a slurry or herbal tea to replenish crop deficiencies (Masson 2018). It is also used in the form of mulch to provide organic elements needed by the crop as it decomposes (Shaikh and Patii 2013).

The infusion also has a slight repellent activity on aphids. The main purpose of comfrey preparations is to improve the soil, so other insect repellent preparations will be preferred if repellent activity is the main concern, but it is an interesting complementary property. Finally, its reputation is also due to its stimulating activity for the microbial flora of the soil and for vegetation in general.

The particular conditions of the spaces under the solar panels led to an intense development, as shown by the abundance of its inflorescences and leafy parts (Figure 3.6).

Comfrey is mainly used for its richness in elements of interest for vegetation, notably potassium and silica. However, its interest as a fertilizer lies especially in its richness in trace elements such as boron, which is often deficient in plants.

Figure 3.6. *Comfrey flowering under solar panels in Boissière (Gard) (source: C. Grison)*

Finally, comfrey contains allantoin. This compound was quickly identified for its healing properties, in healing ulcers, for example, by Dolifus (1936) and Gehdira and Goetz (2012).

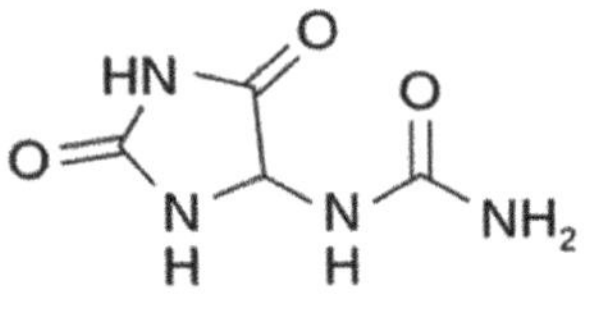

Figure 3.7. *Chemical structure of allantoin*

Its properties are used by the plant to firm up its cells and make it more vigorous. Its proportion in the plant is 9.38 ± 2.72 mg/g dry matter in the leaves and 25.77 ± 17.02 mg/g in the roots (Dresler *et al.* 2017).

3.2.2.4. *Rhubarb,* Rheum rhabarbarum

3.2.2.4.1. Botanical characteristics

Garden rhubarb is a massive, hardy perennial that usually grows in the shade in very moist soil. This hybrid species is the one found in most gardens. This plant prefers cold climates and the optimal pH for its cultivation and propagation is slightly acidic to neutral, between 6.0 and 6.8 (Roggemans and Boxus 1988). It is present in many Eurasian countries. Thanks to its great intolerance to drought, it is more present in the northern half of France and in the countries of eastern Europe. It is also frequently found in gardens in Canada.

It is grown for consumption of its petioles (leaf stems). Its harvest usually begins in the first year of planting and rhubarb can be harvested over three years (Bratsch and Mainville 2005; AgroSemens 2014). Planting density is relatively low because the plant is very massive. In organic cultivation, an area of 1.5 m^2/plant is recommended.

It is very well acclimatized in the shade of solar panels. Light watering by drip is enough for it. Each plant was spaced at least two meters apart with alternating melliferous plants.

Figure 3.8. *Rhubarb plant under solar panels in Boissière (Gard) (source: C. Grison)*

3.2.2.4.2. Use in biodynamic agriculture

Like alder buckthorn, rhubarb is an elicitor for grapevine since it is able to stimulate its defense mechanisms under conditions of attack by herbivores (Sentenac and Thierry 2008). It is appreciated for its fungicidal power against mildew when coupled with copper. Rhubarb acts by necrosing the areas of the leaf infected by mildew. In this way, it prevents the development and sporulation of the fungus.

However, unlike alder buckthorn, rhubarb does not retain the *priming* effect and can be used for preventive purposes (Gindro *et al.* 2007). Thus, it can be used as a preventive treatment on crops, but for a curative treatment of mildew, alder buckthorn will more likely be chosen as it will be more effective.

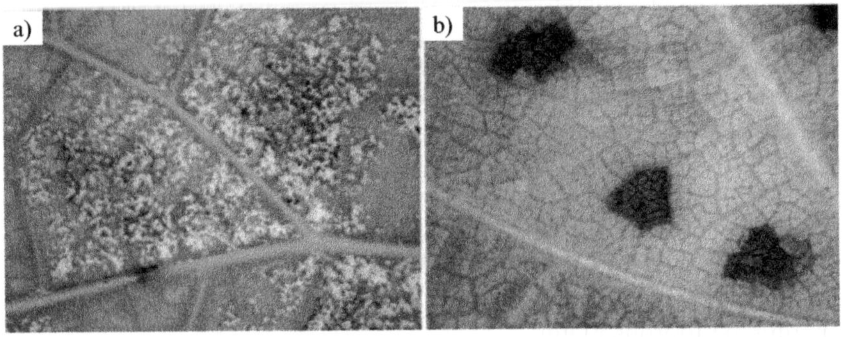

Figure 3.9. *a) Leaf of grapevine infested by downy mildew without treatment (115 sporangia/mm²); b) leaf of grapevine treated with 5% rhubarb extract (no sporulation) (source: Gindro* et al. *(2007))*

Indeed, rhubarb stimulates the dimerization of resveratrol into viniferins that inhibit mildrew spreading (Chastang 2014; Jeandet *et al.* 2020) to control mildew. Its eliciting capacity is less than that of alder buckthorn (viniferin levels about 20 times higher than normal) in case of attack, but its preventive action makes it a very interesting alternative to copper (viniferin levels about 10 times higher than normal if the plant is healthy). If there is no attack, the action of rhubarb extract lasts 10 days, and a drastic drop in viniferin levels is observed from the 11th day.

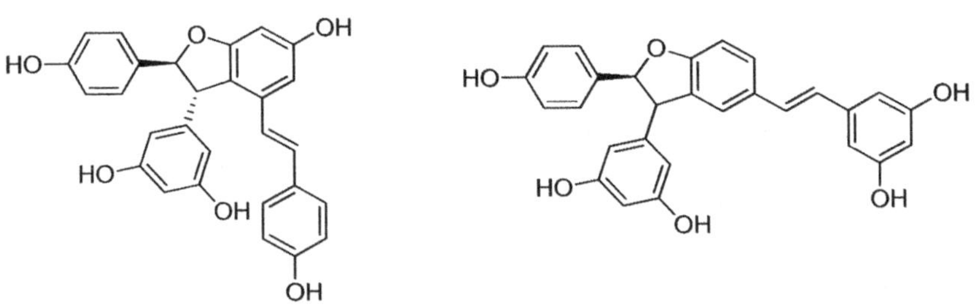

Figure 3.10. *ε-viniferin and δ-viniferin*

The fungicidal capabilities of this plant do not stop at downy mildew, as trials have also been conducted to control fusarium in wheat. Common rhubarb also has insecticidal properties, particularly against aphids, worms and moths in vegetable gardening (ITAB *et al.* 2013; Masson 2018). Active research is underway to better understand these effects. Some chemical data are presented in Appendix 3 of this book. They illustrate the interest of rhubarb in the search for natural active molecules.

3.2.2.4.3. Other interests of rhubarb

Rhubarb is used in traditional Chinese medicine for its numerous therapeutic effects. It has been shown to have a beneficial effect on the liver, spleen and gallbladder. It also has antimicrobial properties (Kosikowska *et al.* 2010).

As previously described, many secondary metabolites present in rhubarb have many interesting medicinal effects. According to the literature, these molecules have a weak cytotoxic effect on healthy cells (Kosikowska *et al.* 2010).

Glucosides from rhubarb have anti-diabetic effects through their metabolic role in the insulin-stimulated glucose transport pathway (Lee and Sohn 2008).

Preparations from rhubarb have already been submitted to the Council of Europe for registration as a basic substance, but these were rejected because it could not be clearly determined that the fungicidal molecules of the plant would not be toxic to non-target organisms and thus present a potential hazard, in particular to soil and water micro-fauna (European Food Safety Authority (EFSA) 2014).

3.2.2.5. *Borage,* Borrago officinalis

3.2.2.5.1. Botanical characteristics

Known for its medicinal and culinary properties, borage has been cultivated in gardens since the Middle Ages (Polese 2006). In addition to being planted, it is regularly found on embankments or roadsides, where it grows spontaneously. It originates from the Middle East, the Mediterranean basin and Asia Minor. It has a deep root system, which makes it a drought-tolerant plant. It can also grow in wet soils if they are drained.

This annual, herbaceous plant is 70 to 100 cm high. It often has a main stem, branched, carrying alternate leaves. However, it can take a rosette form if it is exposed to full light. It presents typical characteristics of adaptation to the hot and dry climates of the Mediterranean basin, such as leaves covered with hairs. It grows preferentially in very sunny areas, as well as tolerating semi-shaded areas. Although its optimum pH is around 6.6, it tolerates a wide range of soil acidity, as it can be found in soils with pH between 4.5 and 8.2.

The cultivation of borage under solar panels was the most problematic. The high mortality rate did not allow the production of leafy parts to be sustained.

3.2.2.5.2 Use in biodynamic agriculture

Like comfrey, borage is a Boraginaceae. It has, globally, the same characteristics as the latter, since it has the same type of chemical components. However, its slurry is less known (and therefore less used) because that of comfrey is more concentrated in trace elements and other elements of interest, which makes it more effective.

Borage is also used as a green fertilizer for soil decontamination. Indeed, it is one of the plants (along with certain species of mustard as well as phacelia) used to reduce the impact of an excessive concentration of metals in the soil. Thus, it has been shown that the presence of borage in a soil highly contaminated with various metals (Pb, Zn, Cu, Cd, Sb) could reduce their concentration and improve soil respiration by phytoremediation (Foucault *et al.* 2013).

Borage makes an interesting amendment in the form of a slurry since it is rich in silica and potassium.

Its allantoin concentration is much lower than that of comfrey (0.83 ± 0.27 mg/g dry matter in roots); hence, it is less used as a fertilizer than the latter (Dresler *et al.* 2017).

Its attractive properties for pollinating insects can be used for vegetable crops. For example, the combination of borage and strawberry plants has been shown to increase strawberry production by 35%. The aesthetic quality of the strawberries is also enhanced (Griffiths-Lee *et al.* 2020).

It is cultivated for pharmaceutical purposes, especially the extraction of GLA (γ-linoleic acid), which is a widespread anti-inflammatory compound. In addition to these anti-inflammatory effects, this compound is also used in the treatment of diabetes, heart disease, arthritis and eczema. Its dry plant yield per hectare is between 1.5 and 2 tons. This yield refers to three or four harvests per year, depending on climatic conditions (Asadi-Samani *et al.* 2014).

3.2.3. *Planting of melliferous species*

Insect pollinators are essential to produce vegetable products and for the reproduction of 80% of plant species. Although domestic bees represent a large part of these pollinators, many species of wild bees, diptera, beetles, etc., play an important role in pollination, each with its own specificity and importance. The melliferous interest of a plant is calculated in relation to the exploitable resources of its flower: pollen (polliniferous plants) and nectar (nectariferous plants) (Van Daele 2011; Moerman *et al.* 2016; Institut technique de l'horticulture *et al.* 2017). Beyond these values that quantify their nutritional quality, they must also be attractive to pollinators.

Within the framework of these first works in ecovoltaics, we chose to complement the plantations of ecological interest with melliferous plants judiciously selected according to the above-mentioned criteria. It turns out that bees are currently in danger, especially with the pressure of the Asian hornet, the Varroa mite and pesticides. Thus, providing them with an abundant and quality food source at a suitable time is a good strategy to revitalize the surrounding ecosystems.

Like the other plants of ecological interest, the melliferous plants planted under the solar panels had to have a high tolerance to shade in order to flower despite the lack of sunlight, a good nutritive quality of pollen and nectar, strong attractive power, as well as flowering spread over the longest possible period. The ideal plant would present a second flowering period, especially outside the summer season.

The flowering periods of the different plant species are outlined below. For each, the nutritional qualities of the pollen and nectar were researched in the scientific literature to guide our choice. However, other parameters such as abundance and number of flowers were also taken into account (Van Daele 2011).

Eight plant species were selected based on these characteristics:

– sweet woodruff (*Galium odoratum* (L.) Scop.);

– bellflower (*Campanula* spp.);

– Scotch broom (*Cytisus scoparius* (L.) Link);

– spotted dead-nettle (*Lamium maculatum* (L.) L.);

– hollyhock (*Alcea rosea* L.);

– Latin American fleabane (*Erigeron karvinskianus* DC.);

– hyssop (*Hyssopus officinalis* L.);

– lemon balm (*Melissa officinalis* L.).

Plants of ecological interest installed in the first stage were also of melliferous interest (alder buckthorn, borage, comfrey) (see below). The months of flowering of the different melliferous plants of interest are grouped in the following diagram, in order to verify the availability of resources for pollinators according to the season.

It is interesting to note that the late flowering of the orchards allowed for a supply of pollen until autumn and that the woodruffs and borage allowed for flowering as early as April. The pollen and nectar qualities of the selected plants are listed in Table 3.3.

The rating in this table ranges from 1 to 5 (5 being excellent and 1 being mediocre) (Institut technique de l'horticulture *et al.* 2017; Van Daele 2011).

Only broom suffered from heat and lack of water in the weeks after planting. *Hyssopus officinalis* and *Melissa officinalis* showed good drought resistance. Shade from the panels and a slightly moist lasagna soil allowed good growth.

	J	F	M	A	M	J	J	A	S	O	N	D
Sweet woodruff (*Galium odoratum*)												
Bellflower (*Campanula* spp)												
Scotch broom (*Cysitisus scoparius*)												
Spotted dead-nettle (*Lamium maculatum*)												
Hollyhock (*Alcea rosea*)												
Latin American fleabane (*Erigeron karvinskianus*)												
Lemon balm (*Melissa officiniais*)												
Hyssop (*Hyssopus officinalis*)												
Alder buckthorn (*Frangula alnus*)												
Borage (*Borago officinalis*)												
Common comfrey (*Symphytum officinale*)												

Table 3.2. *Flowering period of honey plants. For a color version of this table, see www.iste.co.uk/grison/ecology.zip*

Plant	Pollen	Nectar	Confidence index
Sweet woodruff	–	–	–
Bellflower	2	2	1
Scotch broom	3	2	–
Spotted dead-nettle	–	–	–
Hollyhock	4	3	–
Latin American fleabane	–	–	–
Hyssop	2	2	3
Lemon balm	2	1	3
Alder buckthorn	3	1	2
Borage	3	1	3
Comfrey	3	0	2

Table 3.3. *Nectar and pollen quality of selected honey plants*

Figure 3.11. *Scotch broom, Latin American fleabane and alder buckthorn under solar panels (source: C. Grison)*

Figure 3.12. Melissa officinalis, *December 19, 2019 (credit: C. Grison)*

Figure 3.13. Melissa officinalis, *July 2, 2020 (credit: C. Grison)*

Figure 3.14. Hyssopus officinalis, *December 19, 2019 (credit: C. Grison)*

Figure 3.15. Hyssopus officinalis, *July 7, 2020 (credit: C. Grison)*

Figure 3.16. *Most attractive melliferous plants: bellflower, hyssop and lotus (source: C. Grison)*

Species such as bellflower, lotus, Scotch broom and dead-nettle were the most attractive to bees.

This situation led to consideration of the creation of an apiary under the panels, by introducing an ecotype of bees adapted to the region and especially to the area where the solar panels were located, so that they could adapt to this particular environment.

Several arguments in favor of the installation of this apiary were established:

– it would allow good pollination of the flowers present and thus better genetic mixing and better adaptation of the plants under solar panels;

– it would help to indicate and quantify the evolution of the place (whether the bees had enough food or not and which plants they used): bio-control;

– it would reintroduce bees to farmers;

– a conservatory apiary would help to raise local bees that are in danger of extinction.

There was, in addition, an interest in installing an apiary in association with the crops already present on the site: the rhubarb extracts make it possible to protect the beehives from *Varroa destructor* (a parasitic mite of the beehives) via its natural production of oxalic acid.

In order for the bees to be able to forage at the flowers in the spring, it was imperative that the apiary be set up and be in good health as soon as the honey plants began to grow.

However, the study site used (Boissière) presented a major disadvantage, which was the recurrent presence of violent wind gusts. Due to the wind blowing under the panels, the site was incompatible with bees and therefore with the installation of beehives under the panels of this site. No apiaries could be created.

3.2.4. *Installation of regionally endangered species*

The IUCN and the INPN proposed a partnership to the regions that wished to make an inventory of their flora and fauna according to the IUCN criteria. The Occitanie region (and formerly the Languedoc-Roussillon region) has not listed its flora in this way. There is therefore no regional red list.

Article 1 of the decree of October 29, 1997, on the list of protected plant species in Languedoc-Roussillon proposes a list of protected taxa.

However, as this list dates from 1997, it must be considered with caution. The status of many taxa has changed (some no longer exist, scientific names are no longer the same, additional data have been discovered, etc.). It is therefore important not to rely entirely on this list.

The plants protected in the sense of article 1 of this decree cannot be commercialized. It is thus necessary to conduct more thorough research with the conservatories of plants to be able to provide the seed banks.

The selected data were thus brought back to the national level. Indeed, the red list of the vascular flora of metropolitan France was studied in order to extract the plants present in the department of Hérault according to the data of the INPN. Only the plants whose status was "probably extinct", "critically endangered" and "endangered" were studied here. All data are available on the IUCN website (IUCN France *et al.* 2018).

The following plant species have been successfully introduced: *Lotus tetragonobolus* and *Iris xiphium*.

Figure 3.17. *Preparation of endangered plants under solar panels in Boissière (Gard):* Lotus tetragonobolus *and* Iris xiphium *(source: C. Grison)*

3.2.5. *Installation of medicinal species*

Another alternative is to introduce plants of medicinal, aromatic and culinary interest under the solar panels. Indeed, our society tends more and more towards the consumption of so-called "soft" medicines coming from plant extractions. The demand is growing and it would be interesting to be able to cultivate these species to reinforce the offer.

For this alternative, it is advisable to pay particular attention to the legislation, which is very rigorous when it comes to human consumption. We can distinguish two cases in the cultivation of these types of plants:

– producers and processors of plants;

– producers and retailers of plants.

Processors have their own plant processing equipment, such as stills or dryers. Generally, producers with expensive equipment (stills) limit themselves to one or two products (lavender, lavandin, etc.). The producers/gatherers who work mainly with drying are very diversified and favor direct sales on markets. They produce/pick small quantities and often supplement their income with another activity.

Resellers are content to find buyers for their products. They try to produce in bulk to be able to meet the demand of their buyers, who are generally processing laboratories, retailers and producer stores. They are then exempt from all the legislation concerning the sale of their finished products (notably the authorizations to sell plants that appear in the pharmacopoeia without being pharmacists).

When producers decide to sell their products themselves, this implies that these products come from plants on the list of those that may be marketed by persons other than pharmacists (article D4211-11 of the Public Health Code). This list, in addition to the scientific and common name of the plant, specifies the part of the plant that can be sold and in what form (as is, in powder, in solution, etc.). This list was used to find potential plants to install under the solar panels. In order to respect the ecosystems present, only plants present in the region were selected.

In order to test the possibility of their cultivation under solar panels, the grapevine, *Vitis vinifera* (*exalta* and *alfina*), was tested as a model plant. Its growth was regular, but inferior to that of plants in the surrounding fields. This was due to the lack of water and light. This plant is known to treat circulatory or venous disorders, especially during hot weather.

This example is interesting, because grapevine has clearly recognized properties in phytotherapy. The leaves are rich in anthocyanins, flavonoids (quercetin and derivatives) and tannins. Grapes contain pro-anthocyanidolic oligomers, which would have a beneficial effect on microcirculatory manifestations. According to the *Académie nationale de pharmacie* (French National Academy of Pharmacy) (Académie nationale de pharmacie 2016), purified extracts of grapevine can be used in the treatment of functional manifestations of veno-lymphatic insufficiency and in the treatment of lymphedema of the upper limb after radiosurgical treatment of breast cancer.

Grapevine extracts are made up of polyphenolic compounds consisting of several specific flavanic monomer units, which act as inhibitors of numerous enzymatic systems, free radical scavengers, antioxidants, angio-protectors and anti-edematous agents. They are also rich in triglycerides of linoleic acid and in resveratrol, which is used in cosmetics against skin aging.

Figure 3.18. *Growth of* Vitis vinifera *under solar panels in Boissière (Gard) (source: C. Grison)*

3.3. Assessment – issues

This ecovoltaics project is particularly innovative, both in its function and location.

One of the first objectives was to set up ways of thinking and acting in order to integrate a new energy and try to reconcile the energy and ecological challenges by connecting photovoltaic technology and various ecological ambitions. Although the feedback presented is still in its infancy, it is possible to identify positive impacts.

3.3.1. *Positive impact on soil quality*

The first challenge concerned the soil: to succeed in cultivating plant species of various interests on uncultivated soils, not cultivable in their present state. Given the impossibility of tilling the soil, since the photovoltaic installation was already present, we opted for lasagna cultivation, inspired by permaculture. The introduction of local plant waste, acquired nearby, allowed us to create a soil enriched in nutrients and cultivable.

Figure 3.19. *Soil enrichment through the construction of lasagnas under solar panels in Boissière (Gard) (source: L. Cases)*

The presence of the panels over the lasagna probably provided partial protection from water evaporation and allowed some moisture to be retained. In addition to a reduction in watering, the shade certainly enhanced the composting process that gradually decomposes the introduced plant waste. Paradoxically, the wind was a greater obstacle than the heat of the site.

It remains to be verified whether the presence of solar panels limits the day–night thermal shocks to the plants.

We can already conclude that the project of cultivation of plant species of ecological interest has definitively enriched a soil poor in carbon nutrients (thanks to the wood waste) and in nitrogen nutrients by the introduction of green waste. Their progressive decomposition will create a rich and balanced humus, which will benefit the site in the long term.

3.3.2. *Positive impact on faunal biodiversity*

Insects are the most abundant group of animal species. They are essential to the functioning of ecosystems thanks to the numerous interspecific interactions they generate. The services they provide to ecosystems are considerable. For example, it is well-known that pollinating insects are responsible for the reproduction of many plant species, including food species. However, the combined effects of global warming, pollution, deterioration of the environment, loss of diversity, destruction of natural habitats and the proliferation of invasive species constitute a direct threat to insects. Their preservation is however important for the maintenance of the balance of ecosystems. An important question of this work is therefore the following: can the introduction of varied and well-chosen plant species contribute to the development of a significant density and biodiversity of insects? If so, which ones?

In order to establish an initial quantitative assessment in terms of insect numbers and diversity, yellow sticky traps were placed in the experiment and under a control plot not used for planting, as an experimental control. These yellow traps allow the capture of certain specific types of insects (such as flies and other flying insects).

After a week on site and checking weather conditions, the traps were collected and analyzed for insect orders. The insects that could be identified were at least 3 mm in size (Table 3.4).

Crickets, bees, wasps, ants, and aphids, along with non-insect arthropods such as spiders, harvestmen and myriapods, were also observed under the vegetated experimental plot.

a) Traps in the experiment

b) Control panel

Figure 3.20. *Placement of traps on the experiment and on an outside area as a control in Boissière (Gard) (source: L. Cases)*

The conclusion of this first faunal inventory, and particularly in terms of insects, is that the difference between the conventional and experimental plots was mainly in the size of the insects. The introduction of lasagna and plants generated a multitude of very small insects (< 3 mm), which were absent under the non-vegetated panels. This study would merit further analysis, as it is clear that the experiment conducted resulted in the development of many small insects.

	Experimental area		Control zone	
Order	**Number**	**Comment**	**Number**	**Comment**
Diptera	12	Mostly small midges, as well as mosquitoes	7	Midges
Hymenoptera	9	Aphids, small wasps	6	Small wasps
Collembola	1		Not detemined	
Homoptera	1		3	
Coleoptera	n.a.		2	

Table 3.4. *Identification of insects larger than 3 mm*

Finally, the inflorescences of *Lotus tetragonolobus* and *Campanula cochleariifolia* were very attractive to bees and performed well as melliferous species.

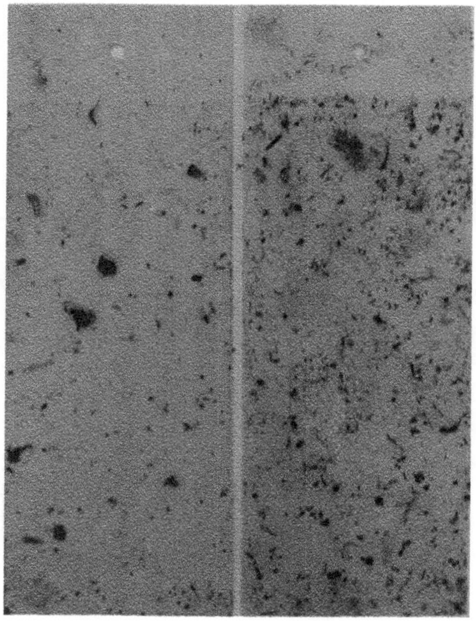

a) b)

Figure 3.21. *Insect traps: difference between control (a) and vegetated (b) plots (source: L. Cases)*

Figure 3.22. Lotus tetragonolobus *flowering (source: L. Cases)*

Figure 3.23. Campanula cochleariifolia *flowering (source: C. Grison)*

3.4. Perspectives

It has been shown from this pioneering work that it is possible under good conditions to introduce plant species of ecological interest: plants for biodynamic agriculture, melliferous plants, medicinal plants and species threatened with extinction.

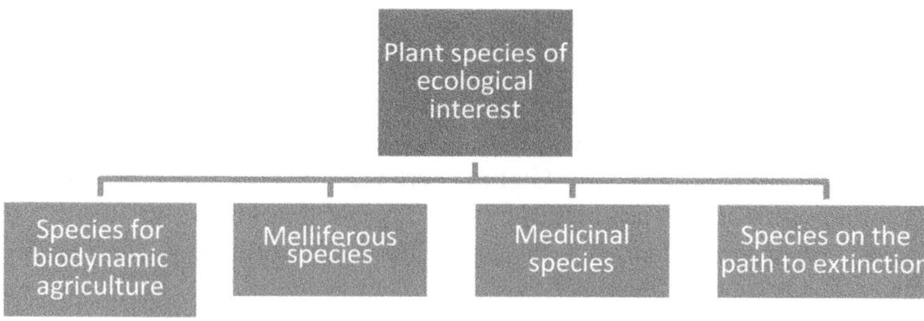

Figure 3.24. *Categories of plant species that can be installed under solar panels*

This first challenge illustrates the possibility of integrating a new and growing energy – photovoltaics – into an exploratory program that is interested in an active and positive ecology. For such initiatives to develop, they must be ecologically, socially and economically sustainable. It is therefore necessary to focus very quickly on the development of each of the sectors addressed. The first factor that can be limiting is regulation. Some thoughts on this subject are presented.

3.4.1. *Regulations for the marketing of plants or plant extracts for use in biodynamic agriculture*

We are interested here in the regulations concerning slurries, decoctions, herbal teas and other plant extracts intended for specific pest control methods in agriculture. According to French law, and in particular the Law on the Future of Agriculture, these products are considered as *Préparations naturelles peu préoccupantes* (PNPPs), that is, Natural Preparations of Little Concern (Assemblée nationale 2014).

Although their use is ancestral and traditional, very few scientific studies have been carried out on the active molecules of the plants from which PNPPs are derived. We lack reliable data on the potential toxicity of these molecules, both at the food and environmental levels. Since the agricultural orientation law of January 5, 2006 (Legifrance 2006), which prohibited their use and commercialization, their legal status has not stopped being modified, and this has led to a marked dissatisfaction among professional producers of these substances, particularly organic farmers. Many actors in the agricultural community have mobilized and declared what they have named "the nettle war" to be able to use the famous nettle manure without being outside the legislation (Aveline and IFV, 2012). Although this is not a proven scientific reference, the Gamm Vert website offers a very complete and synthetic history of the legal texts on PNPPs that allows us to trace the legislation in progress from the beginning of 2006 until today (Gamm Vert 2018).

Even today, there is a succession of laws concerning these products. It is very difficult to know precisely what is authorized and prohibited in the context of a commercial development. Moreover, we must be vigilant to the double legislation which is that of Europe and that of France.

Currently, PNPPs are defined either as basic substances or as biostimulants (Itab 2019). The basic substances are defined by Europe (list available on the Itab website) by article 23 of the EC regulation 1107/2009 as follows:

> A basic substance is an active substance that:
>
> – is not a substance of concern; and
>
> – does not have an inherent capacity to cause endocrine disrupting, neurotoxic or immunotoxic effects; and
>
> – is not predominantly used for plant protection purposes, but nevertheless is useful in plant protection either directly or in a product consisting of the substance and a simple diluent; and
>
> – is not placed on the market as a plant protection product.

> An active substance that meets the criteria of a "foodstuff" as defined in Article 2 of Regulation (EC) No 178/2002 shall be considered as a basic substance.

Biostimulants are defined by French law according to article 50 of the Law on the Future of Agriculture. The plants considered as biostimulants are currently those present in the list of the article D4211-11 of the Code of Public Health, which states that:

> plants or parts of medicinal plants listed in the pharmacopoeia that [...] may, in the form that the list specifies, be sold by persons other than pharmacists (Legifrance 2008).

It is possible to submit applications in order to have new substances registered on the lists concerned. The long duration of the procedures and their cost explain why few applications have been submitted so far, but the legal texts are gradually lightening the procedures concerning the PNPPs to allow easier registration on the lists, including the list of basic substances (EcoPhyto II Plan, Article 253-1 of the Rural and Maritime Fishing Code, service note DGAL/SDQSPV/2019-462, etc.). The approval procedures for a new substance (usable in organic farming) are shown in Figure 3.25 from Itab (this also concerns chemical molecules, referred to here as "active substances", which must also have a marketing authorization (AMM) in France in order to be marketed).

Although nettle manure is currently authorized, many other natural products are not. Products used in human or animal nutrition that can be used as fertilizers or amendments (abiotic natural preparations) are included in the list of Article 50 of the Law on the Future of Agriculture and can be used and sold. On the other hand, natural preparations used for their insecticidal or fungicidal properties, or natural fertilizing preparations based on plants not used in food, must be approved as basic substances.

The files for approval of basic substances are often delegated to Itab, which has a department dedicated to the legislation of new plants. It takes an average of 21 months for the dossiers to be approved by Europe. Over the last seven years, only 53% of basic substances for which applications have been submitted have been approved.

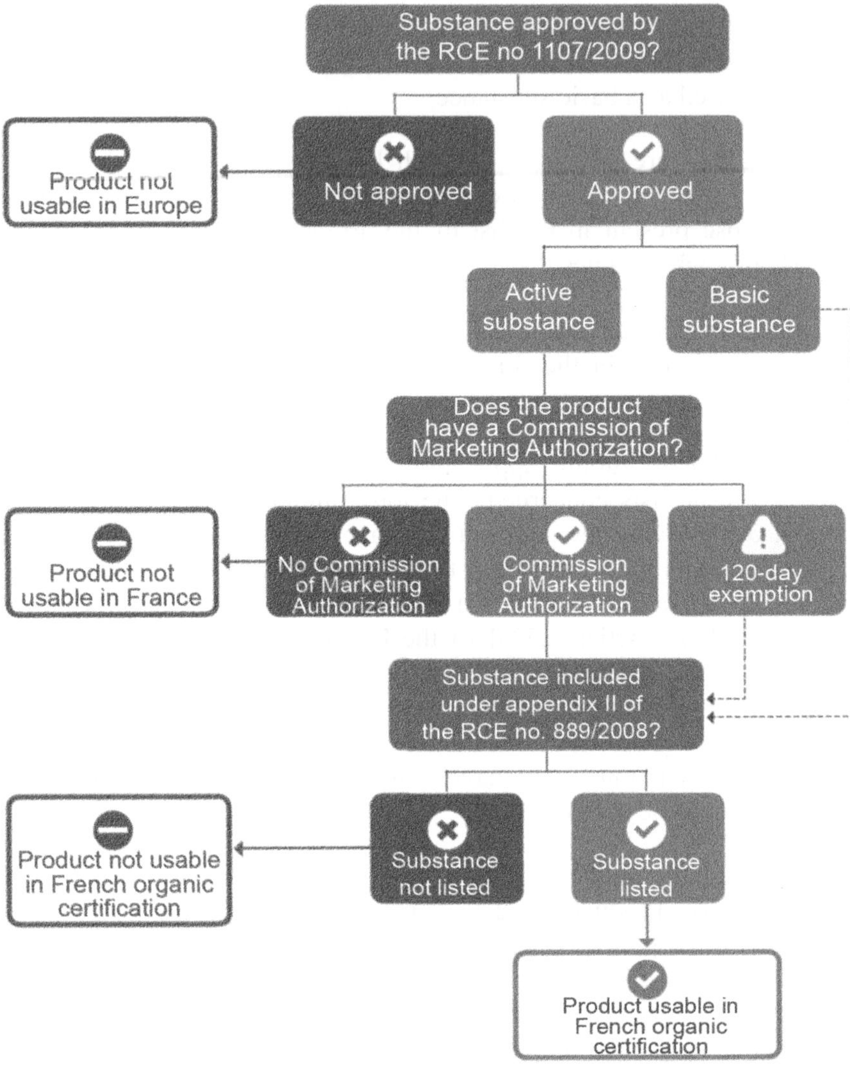

Figure 3.25. *Diagram of legal approval for a new substance (ITAB 2019). For a color version of this figure, see www.iste.co.uk/grison/ecology.zip*

NOTE.– In many EU countries, PNPPs are not considered as plant protection products, and therefore do not have to be listed as basic substances in order to be marketed. This is particularly the case in Germany, Austria, Spain, the United Kingdom and the Netherlands (Agrobio 2012).

Article D4211-11 of the Public Health Code	List of plants and plant extracts with a biostimulant effect that can be marketed
Article L253-1 of the Rural Code	Defines biocontrol products
Article L253-6 of the Rural Code	Provides for the reduction of evaluation times for biocontrol products
Note DGAL/SDQSPV/2019	List of all biocontrol products (active substance + trade name) authorized in organic agriculture
EC Regulation 834/2007	Basic provisions of organic agriculture. Linked to the regulation EC 889/2009
EC Regulation 889/2009	List of authorized products in organic agriculture
EC Regulation 2018/1584	Amendment of the EC regulation 889/2009 by adding new substances to the list
EC Regulation 1107/2009	Regulation only focused on the definition, use and marketing of PNPPs
Law no. 2014-1170 on the Future of Agriculture	Agriculture regulation in general, defines biostimulants

Table 3.5. *Different laws concerning PNPPs*

It is therefore difficult to imagine the development of an economic sector based on the cultivation of plant species for biodynamic agriculture. This sector clearly lacks the scientific knowledge to specify under what conditions plant extracts can be used without risk, in order to replace the phytosanitary chemicals that are so problematic. The marketing of plant extracts for herbal medicine is more mature and constitutes a possible outlet.

Finally, the cultivation of melliferous species and the establishment of seed banks for endangered plant species do not have a financial value but represent an opportunity for the preservation of biodiversity. Why can't we imagine that the financial revenues from the sale of electrical energy could finance such large-scale programs? Wouldn't this be a concept of an energy and ecological transition?

Appendices

Appendix 1

Secondary Metabolites and Defense Molecules of Eagle Fern

A1.1. Prunasin

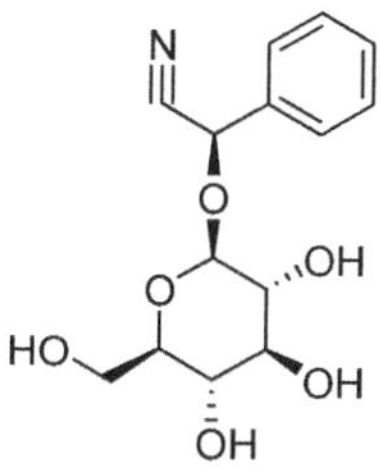

Figure A1.1. *Molecular structure of Prunasin*

Prunasin is a cyanogenic glycoside. A cyanogenic glycoside is not toxic in itself, since this molecule is stable. It is a defense molecule against herbivore attacks and the plant releases hydrogen cyanide (HCN) only after lysis and rupture of tissues, whether by herbivores, by fungal attacks or by mechanical means.

This release of hydrogen cyanide occurs in two stages. First comes the hydrolysis of the cyanogenic glycoside, prunasin, into carbohydrate and cyanohydrin. Then comes the decomposition of cyanohydrin into HCN and the corresponding carbonyl compound.

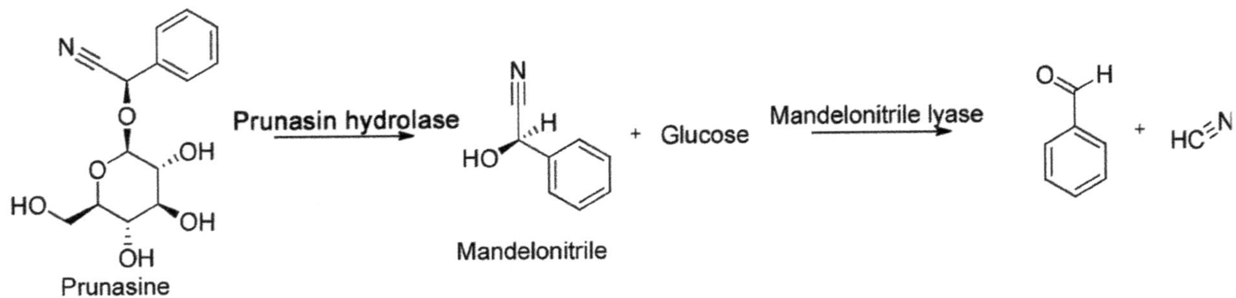

Figure A1.2. *Mechanism of prunasin degradation (Chaouali 2013)*

Glycoside cleavage requires the enzymatic action of a β-glucosidase, while cyanohydrin decomposition can occur spontaneously or enzymatically via a mandelonitrile lyase (MDL). The decomposition reaction is slow under acidic conditions and rapid under alkaline conditions (Chaouali 2013).

This first step does not occur if the plant is intact and thus there is no release of HCN. As explained earlier, hydrolysis of cyanogenic glycoside occurs at a significant rate only after the plant has been deteriorated. Indeed, it is accepted that glycosides and their catabolic enzymes are localized in separate compartments in the intact plant. This is called the principle of compartmentalization, which protects the plant from its own toxic compounds. It is only after the rupture of the plant tissue of cyanogenic plants that the compartmental separation between the enzymes and their substrates breaks down, allowing the catabolic enzymes to access the cyanogenic glycosides (Wajant *et al.* 1994).

Again, the concentration of prunasin in the leaves varies with the stage of development of the plant. The concentration of prunasin in leaves ranges from 10.4 to 61.3 mg/g of fresh plant (Vetter 2009).

One mole of hydrogen cyanide is released per mole of prunasin decomposed. Since fresh eagle fern contains between 10.4 and 61.3 mg of prunasin per gram of fresh leaf, between 0.95 and 5.61 mg of hydrogen cyanide is generated per gram of fresh leaf.

The lethal dose of hydrogen cyanide is 2 mg per kilogram of body weight for livestock (Majak and Cheng 1987). For humans, this dose is between 0.5 and 3 mg of hydrogen cyanide per kilogram of body mass.

The presence of prunasin in ferns is a protection mechanism of the plant against insects and herbivores. The insecticidal effect generated is well-known. However, no herbivore death has been described due to the presence of cyanide in the plant; it only generates a bitter taste for the latter (Alonso-Amelot 2002) and thus often avoidance.

A1.2. Ptaquiloside (also called aquilide A)

The mechanistic study of ptaquiloside degradation and the interaction of its metabolites determine at which stage of maturity the fern should be optimally used.

In a basic medium, ptaquiloside forms an aglycone ptaquilosine, releasing a D-(+)-glucose molecule. Depending on the conditions in which the molecule is present, thus depending on the biological environment, two mechanistic routes are possible. In a basic medium, as is the case in many biological fluids, a very unstable dienone unit is formed. This unit is extremely reactive due to the surrounding geometry of the cyclopropane present.

The formed dienone can lead to the formation of pterosine B in an acidic or basic medium with pH above 8. Pterosine B has no known toxicity since this molecule is stable.

Indeed, the dienone intermediate acts as the defense molecule by selectively alkylating nucleic acids through a Mickael-type addition. More precisely, the dienone cleaves DNA at the guanine and adenine nucleic acids. There is a selectivity according to the incubation time of the molecule in the presence of DNA (selectively cleaves adenine after 25 h and guanine at 130 h) (Yamada *et al.* 1998).

The eagle fern produces other molecules with a similar structure, like caudatoside and methylptaquiloside.

The mechanism of degradation of ptaquiloside analogues is the same as for ptaquiloside, and therefore the same interactions with DNA are generated (Potter and Baird 2000).

The presence of sesquiterpene-like compounds such as ptaquilosides is higher in young fronds (Chen *et al.* 2008; Vetter 2009). The maximum concentration observed in fronds was obtained in May in Europe (Rasmussen *et al.* 2013). The concentration of ptaquiloside varies with leaf development and can reach concentrations of 40,000 µg/g. In fact, ptaquiloside is transferred to the rhizomes at the end of plant growth (Rasmussen *et al.* 2015). Another factor that influences the concentration of ptaquiloside in fronds is rainfall. When it rains, the concentration decreases in the leaves (O'Driscoll *et al.* 2016). The average concentration is between 500 and 5,000 µg/g (Aranha *et al.* 2019).

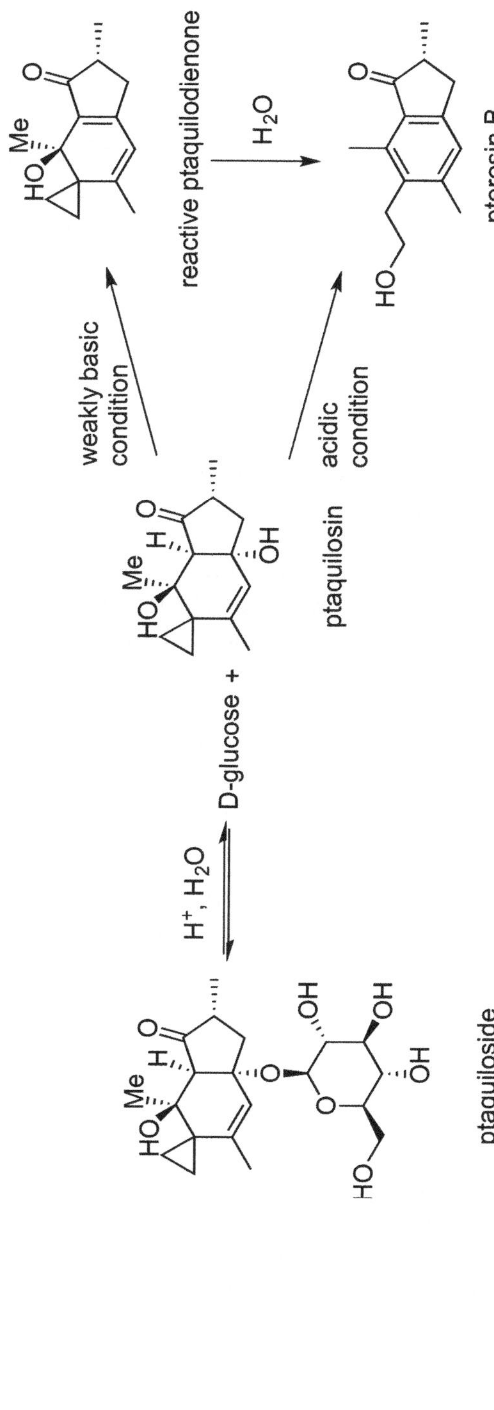

Figure A1.3. *Degradation of ptaquiloside*

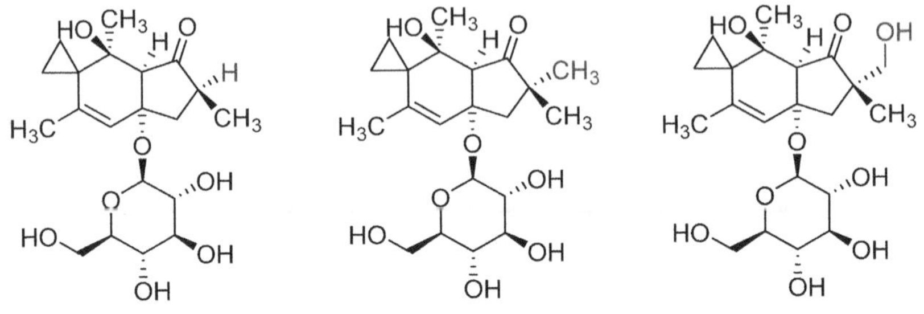

Figure A1.4. *Ptaquiloside and derivatives – defense molecules present in Eagle fern*

A1.3. Thiaminase I (antithiamine)

Thiaminases are enzymes that degrade thiamine, that is, vitamin B1, through the following mechanisms.

Thiaminase I degrades thiamine to form a thiazole and a pyridine derivative. However, thiaminase I needs a co-substrate, which is usually a base, but can also be a nucleophile, such as sulfinic acid, or a thiol.

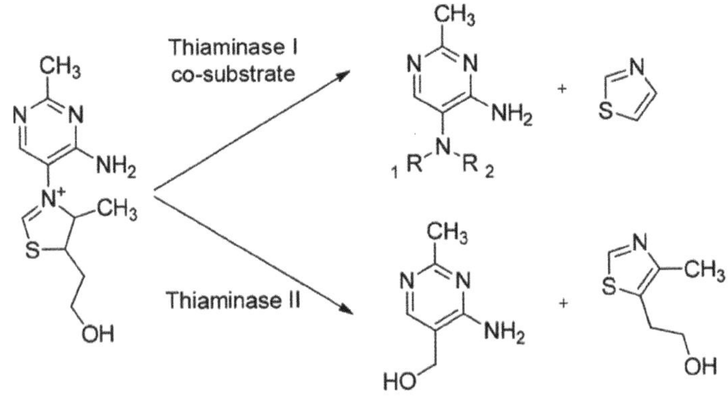

Figure A1.5. *Mechanism of action of thiaminases*

Thiamine is a metabolic precursor of thiamine pyrophosphate (TPP), an essential coenzyme for some decarboxylases. This vitamin is involved in the Krebs cycle, in the transformation of carbohydrates into energy. It allows the transformation of pyruvate produced by glycolysis.

In animals, thiamine is a water-soluble vitamin of the B vitamin family that they must find in their diet through plants. It is also produced by bacteria or fungi.

In humans, a dietary deficiency of vitamin B1 causes beriberi and can also cause Gayet–Wernicke encephalopathy.

Thiaminase is clearly more present in rhizomes, especially in the fall, around September. In young fronds, there is a maximum concentration in May in Europe (Figure A1.6).

Thiaminase is deactivated during steaming of the eagle fern; it is a thermolabile molecule. When heat treatment is performed at low temperatures (30–40°C), not all molecules are degraded, whereas at 100°C, all are (Evans 1976).

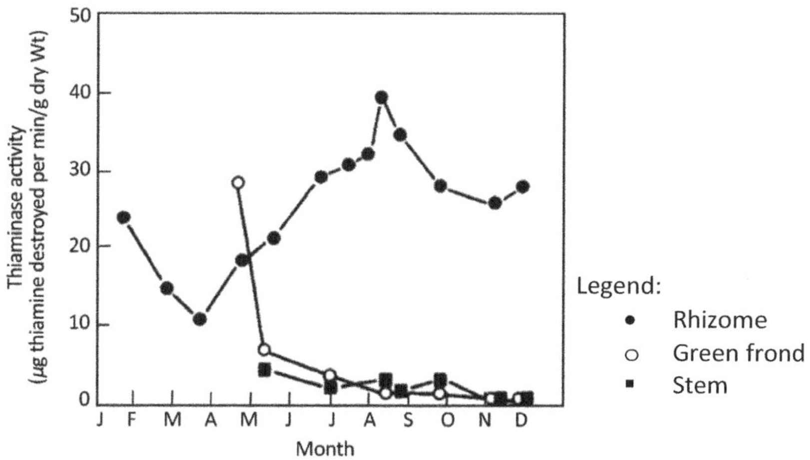

Figure A1.6. *Seasonal variation of thiaminase concentration in Eagle fern (Evans 1976)*

A1.4. Other metabolites

Other interesting and non-toxic components are also present in the eagle fern, such as the β-sitosterol (Chen *et al.* 2008) which is a steroid. This molecule is used to treat prostate hypertrophy, a benign disease in the elderly. It has also been studied for its protective role against colon, prostate and breast cancer (Awad *et al.* 2000).

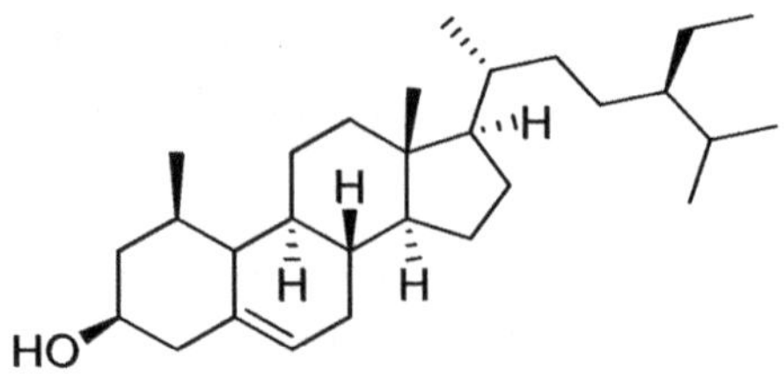

Figure A1.7. *Structure of β-sitosterol*

Appendix 2

Secondary Metabolites and Defense Molecules of Alder Buckthorn

The molecules responsible for this elicitor effect are anthraquinones, present in large quantities in the bark of alder buckthorn. They are known for their phytotoxic properties against many cancer cells.

Anthraquinones are organic pigments derived from the basic structure of 9,10-anthracenedione (Gonçalves *et al.* 2018), as shown in Figure A2.1.

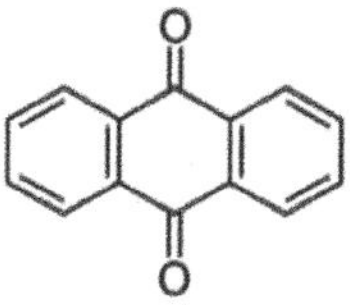

Figure A2.1. *Chemical structure of an anthraquinone*

The anthraquinones present in the bark of alder buckthorn are mainly emodin, chrysopanol and physcione, as shown in Figure A2.2. (Gonçalves *et al.* 2018).

Treatment of mildew-infected grapevine with anthraquinones from alder buckthorn stimulates the synthesis of defense molecules in grapevine. At sites of infection, grapevine defense results in the production of specific proteins, newly synthesized metabolites (phytoalexins) and complex constituent substances (callose) capable of limiting or inhibiting downy mildew development (Gindro *et al.* 2007). Stilbene phytoalexins have a biocidal effect on the fungus (notably ε- and δ-viniferin). Other active downy mildew defense molecules are resveratrol and pikeid (Gindro *et al.* 2007). The chemical structures of grapevine defense molecules are shown in Figure A2.3.

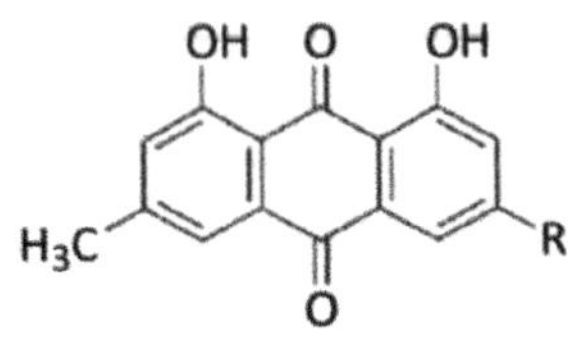

(1) Chrysophanol, R = H
(2) Physcione, R = $-OCH_3$
(3) Emodin, R = -OH

Figure A2.2. *Chemical structures of chrysophanol (1), physcione (2) and emodin (3)*

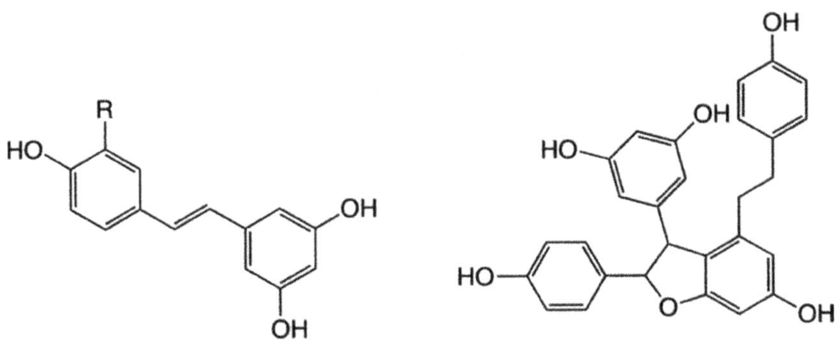

R = H, resveratrol
R = OH, pieceatannol

δ-viniferin

Figure A2.3. *Molecular structures of grapevine defense mechanisms*

Appendix 3

Secondary Metabolites and Defense Molecules of Rhubarb

A3.1. Carboxylic acids

The concentration of oxalic acid, usually between 600 and 5,000 mg/kg, is highly variable and depends on many parameters (Tureckova 2001). This concentration also varies according to the part of the rhubarb studied, and it is maximal in the leaves.

The presence of gallic acid has been reported in many studies (Kalisz *et al.* 2020). In addition, this acid is produced during the degradation of many of the compounds presented below.

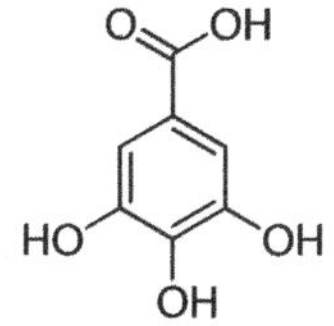

Figure A3.1. *Gallic acid*

Gallic acid induces a massive production of stilbenes over a short period of time, insufficient to inhibit the development of mildew, but it allows interesting protection through its biocide effect.

The presence of malic acid, citric acid and fumaric acid has also been noted (Kalisz *et al.* 2020).

A3.2. Minerals

Rhubarb contains high concentrations of certain minerals, such as potassium (up to 4,000 mg/L) and magnesium. On the other hand, there is no calcium in this plant since this element precipitates in the presence of oxalic acid, naturally present in rhubarb (Will and Dietrich 2013).

A3.3. Anthraquinones

Several anthraquinones are present in rhubarb (Table A3.1) along with their corresponding glucosides (Zheng 2013). *Rheum rhabarbarum* contains a total of 18.1 ± 0.8 mg/g of anthraquinones (Kosikowska *et al.* 2010). Studies have demonstrated the interest of many anthraquinones for their anticancer actions. Different molecules can inhibit cancer proliferation without having the same mechanisms of action, which makes their therapeutic potential particularly interesting (Huang *et al.* 2007).

Molecule	Effects described in the literature
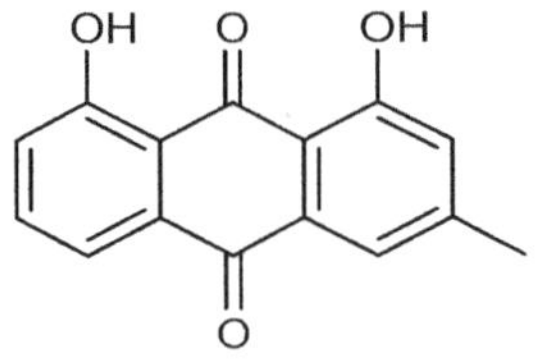 Chrysophanol	Anticancer, antiviral, antidiabetic, anti-inflammatory, anti-protozoal, hypolipidemic, hepatoprotective, neuroprotective, anti-ulcer and anti-obesity effects
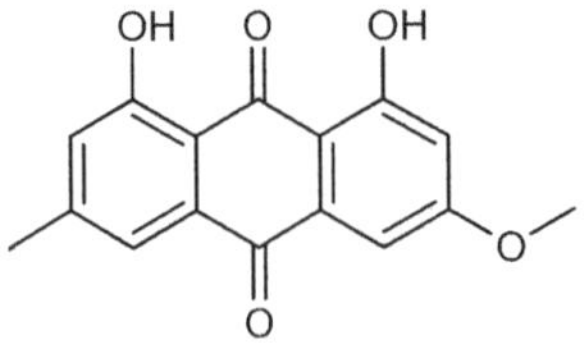 Physcion	Anti-inflammatory, hepatoprotective, antifungal and anticancer effects
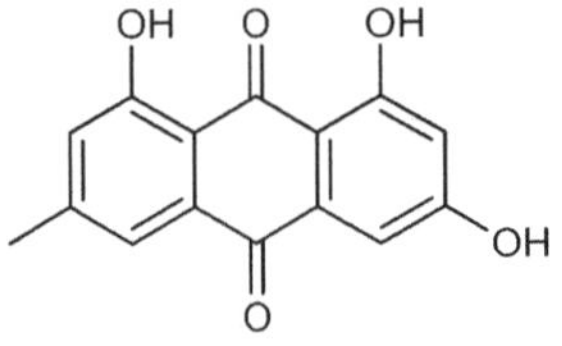 Emodin	Antibacterial, anti-inflammatory, antiviral, anti-ulcerogenic, anticancer, immunosuppressive and chemopreventive effects Inhibits the proliferation of cancer cells (ovarian, breast, lung, liver and prostate cancer) (Hsi and Chung 2012)

Molecule	Effects described in the literature
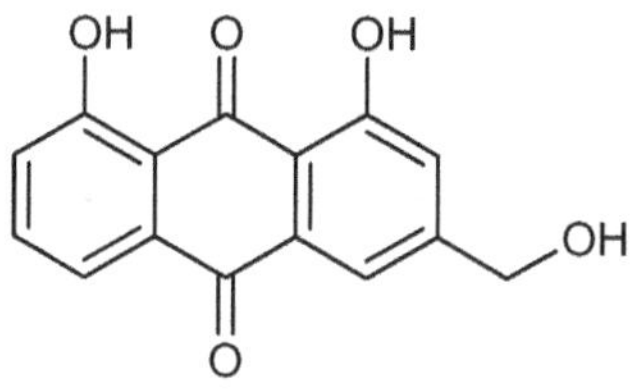 Aloe-emodin	Antineoplastic agent (prevents the spread and maturation of malignant cells); this effect is due to the multi-channel mechanisms of this molecule, including cell cycle disruption; induction of apoptosis and anti-metastatic and anti-angiogenic mechanisms; and enhancement of immune function (Chen *et al.* 2014)
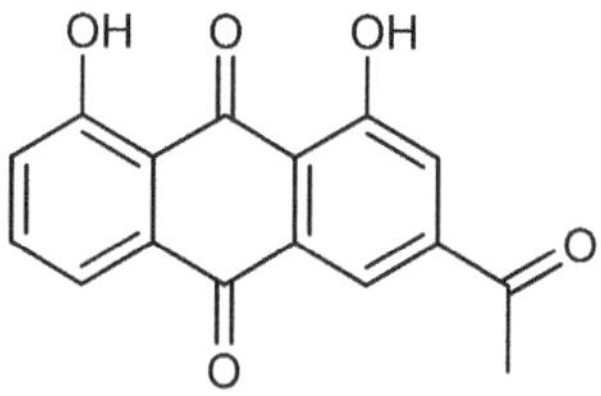 Rhein	Anti-inflammatory, antitumor, antioxidant, antifibrotic, hepatoprotective and nephroprotective effects (Sun *et al.* 2016)

Table A3.1. *Anthraquinones present in* Rheum rhabarbarum

A3.4. Stilbenes

Rhubarb is known for its antioxidant activities which can be linked to the presence of stilbenes, including resveratrol, rhaponticin and piceatannol, in its rhizomes (He *et al.* 2019). Other stilbenes are also present and presented in Table A3.2. They can also be found associated with carbohydrates (Zheng *et al.* 2013).

Stilbenes are secondary metabolites. Their role is to provide protection and resistance in plants against diseases. Their synthesis is often a response to a pathogen attack. These molecules are thus called phytoalexins (Gabaston 2018).

Resveratrol is the best-known stilbene in the world. It is also the most studied, although the others are of increasing interest to the scientific community and are therefore being studied more and more.

Molecule	Effects described in the literature
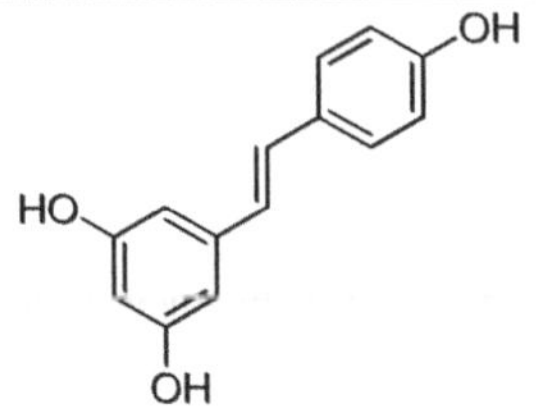 Resveratrol	The most well-known antioxidant due to its presence in grapevine; it has various known biological activities, such as antifungal, anti-inflammatory, antioxidant and antidiabetic activities, and provides cardiovascular and anticancer benefits (Nieder *et al.* 2006; Piotrowska *et al.* 2012; Dong *et al.* 2015)
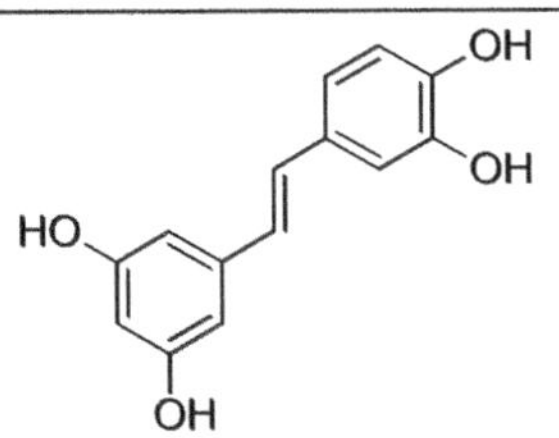 Piceatannol	Antioxidant (more effective than resveratrol), antiparasitic and antibacterial properties prevent the proliferation of many cancers (breast, prostate, colon and melanoma) (Piotrowska *et al.* 2012)
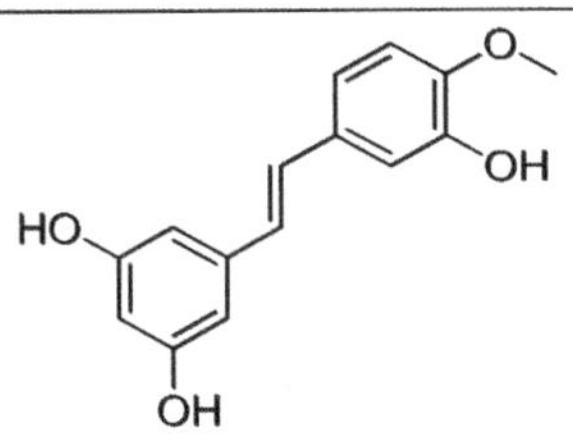 Rhapontigenin	Anti-allergic, purgative, anticoagulant, anti-thrombotic, tyrosine kinase inhibitor (cancer treatment) and hypoglycemic effects (Zhang *et al.* 2007)
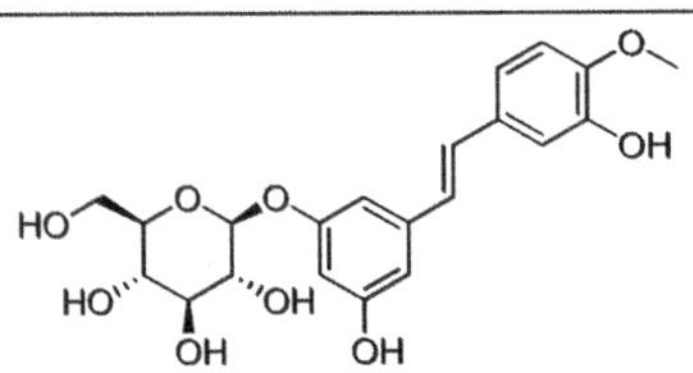 Rhaponticin	Anti-allergic, purgative, anticoagulant, anti-thrombotic, tyrosine kinase inhibitor (cancer treatment) and hypoglycemic effects (Zhang *et al.* 2007)

Table A3.2. *Main stilbenes present in* Rheum rhabarbarum

A3.5. Flavonoids

Finally, other polyphenolic compounds are present in rhubarb; these are flavonoid compounds, as presented in Table A3.3 (Zhenget *et al.* 2013).

<table>
<tr><th>Molecule</th><th>Effects described in the literature</th></tr>
<tr><td>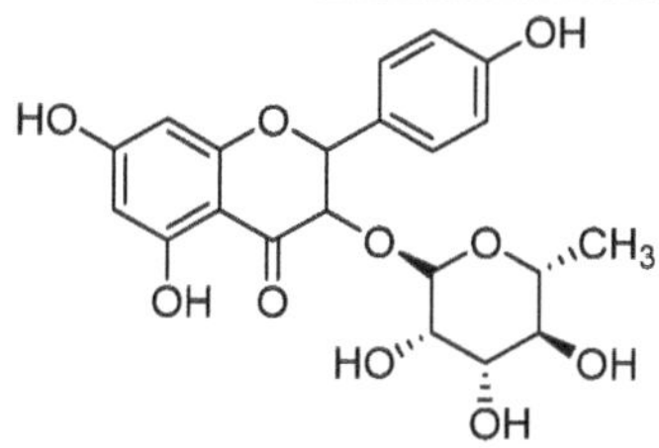
Kaempferol-3-O-rhamnoside</td><td rowspan="2">Inhibit the proliferation of cancer cells and induce their apoptosis (Diantini et al. 2012)</td></tr>
<tr><td>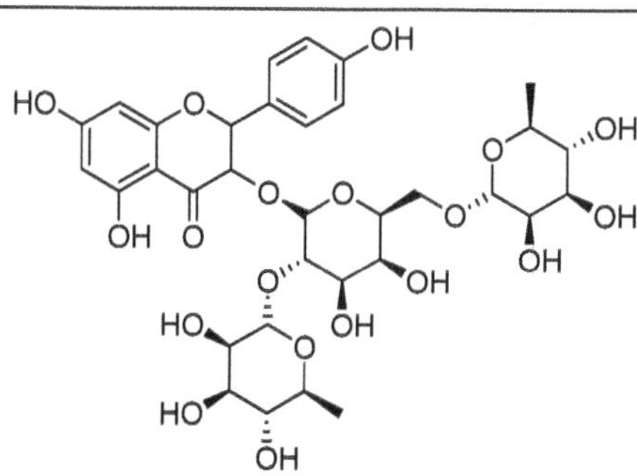
Kaempferol-3-O-(2',6'-di-O-rhamnopyranosyl)ß-D-galactopyra-noside</td></tr>
</table>

Table A3.3. *Flavonoids present in* Rheum rhabarbarum

References

Académie Nationale de Pharmacie (2016). Vigne [Online]. Available at: http://dictionnaire.acadpharm.org/w/Vigne.

Acthuel (2020). Agrivoltaïsme : recensement des principales applications [Online]. Available at: https://www.acthuel.com/fr/agrivoltaisme.

Actualité Parlement européen (2018). Réduction des émissions de gaz à effet de serre dans l'UE : objectifs pour 2030 [Online]. Available at: https://www.europarl.europa.eu/news/fr/headlines/society/20180208STO97442/reduction-des-emissions-de-gaz-a-effet-de-serre-dans-l-ue-objectifs-pour-2030.

ADEME (2013). Estimations d'après méthode ClimAgri®. Report, ADEME.

ADEME (2015). Maîtriser l'énergie en agriculture : un objectif environnemental et économique. Report, ADEME.

ADEME (2017). Feuille de route du photovoltaïsme. Roadmap, ADEME.

ADEME (2018a). Intégration des énergies renouvelables et de récupération dans l'industrie – Fiche technique Solaire photovoltaïque. Data sheet, ADEME.

ADEME (2018b). L'autoconsommation d'électricité d'origine photovoltaïque. Report, ADEME.

ADEME (2018c). Centrale photovoltaïque en autoconsommation – retour d'expérience. Report, ADEME.

ADEME (2019a). Évaluation du gisement relatif aux zones délaissées et artificialisées propices à l'implantation de centrales photovoltaïques. Report, ADEME.

ADEME (2019b). Avis de l'ADEME sur le Solaire photovoltaïque. Note, ADEME.

Agrienergie® Project (2010). Akuo [Online]. Available at: https://www.akuoenergy.com/fr/pierrefonds.

Agrobio 47 (2012). Guide technique Plantes à parfum, aromatiques et médicinales (PPAM) en bio. Guide, Agrobio, Villeneuve-sur-Lot.

AgroSemens (2014). AGROSEMENS - Catalogue 2014 [Online]. Available at: https://issuu.com/arnaudmb/docs/catalogue_2014_complet_moyenne_def.

Alonso-Amelot, M.E. (2002). The chemistry and toxicology of bioactive compounds in bracken fern (*Pteridium* ssp), with special reference to chemical ecology and carcinogenesis. *Stud. Nat. Prod. Chem.*, 26, 685–739.

Aranha, P.C., Dos, R., Rasmussen, L.H., Wolf-Jäckel, G.A., Jensen, H.M.E., Hansen, H.C.B., Friis, C. (2019). Fate of ptaquiloside – a bracken fern toxin – in cattle. *PLOS One*, 14, e0218628 [Online]. Available at: https://doi.org/10.1371/journal.pone.0218628.

Asadi-Samani, M., Bahmani, M., Rafieian-Kopaei, M. (2014). The chemical composition, botanical characteristic and biological activities of *Borago officinalis*: A review. *Asian Pac. J. Trop. Med.*, 7, S22–S28 [Online]. Available at: https://doi.org/10.1016/S1995-7645(14)60199-1.

ASPRO PNPP (2019). Association pour la promotion des préparations naturelles peu réoccupantes [Online]. Available at: https://www.aspro-pnpp.org/wp-content/uploads/2019/05/Lettre-ouverte-%C3%A0-Monsieur-le-Ministre-de-lAgriculture.pdf.

Assemblée nationale (2014). Loi n° 2014–1170. Avenir pour l'agriculture, l'alimentation et la forêt.

Assemblée nationale (2020). Réacteurs en exploitation : se préparer au démantèlement à grande échelle. LCP [Online]. Available at: https://www.irsn.fr/FR/connaissances/Installations_nucleaires/demantelement/demantelement-centrales-installations-nucleaires-EDF-recherche-militaire/Pages/4-preparer-demantelement-grand-echelle-centrales-installations-nucleaires.aspx#.X0_ZVnkzbIV.

Association PV Cycle (n.d.). Recyclage [Online]. Available at: https://pvcycle.fr/recyclage/.

Association technique viticole (2013). Bulletin technique viticulture no. 32. Bulletin, Association technique viticole.

Aveline, N. (2012). Lutte contre les bio-agresseurs : le point sur les produits alternatifs. *IFV*, 14.

Awad, A.B., Chan, K.C., Downie, A.C., Fink, C.S. (2000). Peanuts as a source of β-sitosterol, a sterol with anticancer properties. *Nutr. Cancer*, 36, 238–241 [Online]. Available at: https://doi.org/10.1207/S15327914NC3602_14.

Barron-Gafford, G.A., Pavao-Zuckerman, M.A., Minor, R.L., Sutter, L.F., Barnett-Moreno, I., Blackett, D.T., Thompson, M., Dimond, K., Gerlak, A.K., Nabhan, G.P., Macknick, J.E. (2019). Agrivoltaics provide mutual benefits across the food-energy-water nexus in drylands. *Nat. Sustain.*, 2, 848–855 [Online]. Available at: https://doi.org/10.1038/s41893-019-0364-5.

Belkhayat, M. (2017). Les fours de cristallisation du silicium photovoltaïque. *Encyclopédie de l'énergie* [Online]. Available at: https://www.encyclopedie-energie.org/les-fours-de-cristallisation-du-silicium-photovoltaique/.

Biotope Agence Océan Indien (2013). Projet de ferme photovoltaïque "Les Cèdres" – Commune de L'Étang Salé [Online]. Available at: https://www.eib.org/attachments/registers/50267656.pdf.

Boetsch, G. and Späni, A. (2013). *La grande muraille verte : des arbres contre le désert*. Privat, Toulouse.

Boulanger, V. (2020). Modules verticaux : l'option agrivoltaïque. *Journal du photovoltaïque*, 34, 30–31.

Boxenbaum, E., Laurent, B., Lacoste, A. (2013). *Nouvelles énergies pour la ville du futur*. Presses des Mines, Paris.

Bratsch, A. and Mainville, D. (2005). *Specialty crop profile: Rhubarb*. Document, Virginia Tech, 1–6.

Cadmium (2010). Fiche résumée écotoxicochimique [Online]. Available at: http://www.atctoxicologie.fr/images/Dossier/FRTEC/cdte_frt ec_17.pdf.

Cassoret, B. (2020). *Transition énergétique : ces vérités qui dérangent !* De Boeck Supérieur, Paris.

Centre national de ressources photovoltaïques (2020). Photovoltaique info [Online]. Available at: https://photovoltaique.info/fr/.

Chambre d'agriculture de l'Aude (2019). Deux sites pilotes d'agrinergie dans l'Aude et le Gard [Online]. Available at: https://www.midilibre.fr/2019/06/07/deux-sites-pilotes-dagrinergie-dans-laude-et-le-gard,8243891.php and https://www.lindependant.fr/2019/06/07/deux-sites-pilotes-dagrinergie-dans-laude-et-legard,8243837.php.

Chambre régionale d'agriculture des Pays de la Loire (2010). Réussir son bâtiment agricole avec du photovoltaïque. Report, Chambre régionale d'agriculture des Pays de la Loire.

Chaouali, N. (2013). Intoxication par les plantes cyanogènes. Thesis, University of Monastir, Monastir.

Chastang, T. (2014). Étude de la synthèse du resvératrol et de ses dérivés (viniférines) par des suspensions de cellules de vigne et optimisation de la production en bioréacteur. Thesis, École centrale de Paris, Paris.

Chen, Y., Yong, Z., Yuanjie, H., Liquin, W., Zhongtao, D., Ying, L., Jihua, W. (2008). Isolation of 5-hydroxypyrrolidin-2-one and other constituents from the young fronds of *Pteridium aquilinum*. *J. Nat. Med.*, 62(3), 358–359.

Chen, R., Zhang, J., Hu, Y., Wang, S., Chen, M., Wang, Y. (2014). Potential Antineoplastic effects of aloe-emodin: A comprehensive review. *Am. J. Chin. Med.*, 42, 275–288 [Online]. Available at: https://doi.org/10.1142/S0192415X14500189.

Ciel et Terre (n.d.). Notre concept de solaire flottant international [Online]. Available at: https://www.ciel-et-terre.net/fr/presence-mondiale/.

Combe, M. (2020). Le photovoltaïque s'invite dans le monde agricole. *Techniques de l'ingénieur* [Online]. Available at: https://www.techniques-ingenieur.fr/actualite/articles/le-photovoltaique-sinvite-dans-le-monde-agricole-76217/.

Commission européenne (2016). Accord de Paris. Action pour le Climat [Online]. Available at: https://ec.europa.eu/clima/policies/international/negotiations/paris_fr.

Commission européenne (2019). Un pacte vert pour l'Europe [Online]. Available at: https://ec.europa.eu/info/strategy/priorities-2019-2024/european-green-deal_fr.

Crampon, C. (2019). Confusion sur les chiffres d'émissions de CO_2 du nucléaire, toutes les infos. *Revue Générale Nucléaire*, SFEN [Online]. Available at: https://www.sfen.org/rgn/confusion-chiffres-emissions-CO_2-nucleaire.

Davies, J., Poulsen, L., Schulte-Herbrüggen, B., Mackinnon, K., Crawhall, N., Henwood, W.D., Dudley, N., Smith, J., Gudka, M. (2012). Conservation de la biodiversité des zones arides. Report, UICN, PNUE-WCMC, CNULCD.

Décisions durables (2017). Japon : des champignons sous les panneaux solaires [Online]. Available at: https://www.decisionsdurables.com/japon-des-champignons-sous-les-panneaux-solaires/.

Delarue, E. (2019). Réussir fruits légumes FLD [Online]. Available at: https://www.reussir.fr/fruits-legumes/experimentation-la-pugere-sort-de-la-zone-dombre.

Deleporte, E. (2017). Les pérovskites hybrides : des excitons sur mesure pour l'optoélectronique. *Photoniques*, 40–43 [Online]. Available at: https://doi.org/10.1051/photon/20178740.

Diantini, A., Subarnas, A., Lestari, K., Halimah, E., Susilawati, Y., Supriyatna, S., Julaeha, E., Achmad, T.H., Suradji, E.W., Yamazaki, C., Kobayashi, K., Koyama, H., Abdulah, R. (2012). Kaempferol-3-O-rhamnoside isolated from the leaves of *Schima wallichii* Korth. inhibits MCF-7 breast cancer cell proliferation through activation of the caspase cascade pathway. *Oncol. Lett.*, 3, 1069–1072 [Online]. Available at: https://doi.org/10.3892/ol.2012.596.

Direction départementale des territoires et de la mer de l'Hérault (2011). Guide méthodologique sur le photovoltaïque dans l'Hérault [Online]. Available at: http://www.herault.gouv.fr/content/download/1390/8356/file/guide_photovoltaique2011.pdf.

Dollfus, R.P. (1936). Vieux remèdes – larves d'asticots et protidothérapie. *Terre Vie*, 4, 232.

Dong, G., Lee, Y., Jeong, J.H., Zhao, H.-Y., Jeon, R., Lee, H.J., Ryu, J.-H. (2015). Stilbenoids from *Rheum undulatum* protect hepatocytes against oxidative stress through AMPK activation: Stilbenoids protect hepatocytes against oxidative stress. *Phytother. Res.*, 29, 1605–1609 [Online]. Available at: https://doi.org/10.1002/ptr.5442.

Dresler, S., Szymczak, G., Wójcik, M. (2017). Comparison of some secondary metabolite content in the seventeen species of the Boraginaceae family. *Pharm. Biol.*, 55, 691–695 [Online]. Available at: https://doi.org/10.1080/13880209.2016.1265986.

Dumas, Y. (2002). Que savons-nous de la fougère aigle ? *Rev. For. Fr.*, 4, 357–374.

Dupin, L. (2016). L'usine énergie, énergies renouvelables, photovoltaïque, solaire. *L'Usine Nouvelle*.

Dupraz, C., Marrou, H., Talbot, G., Dufour, L., Nogier, A., Ferard, Y. (2011). Combining solar photovoltaic panels and food crops for optimising land use: Towards new agrivoltaic schemes. *Renew. Energy*, 36, 2725–2732 [Online]. Available at: https://doi.org/10.1016/j.renene.2011.03.005.

Elamri, Y. (2017). Bilan hydrique et développement de culture sous panneaux photovoltaïques dynamiques : de la modélisation à l'évaluation de solutions agrivoltaïques. Thesis, University of Montpellier, Montpellier [Online]. Available at: http://www.g-eau.fr/index.php/fr/productions/theses/item/677-optimisation-de-la-strategie-d-irrigation-et-de-la-production-vegetale-sous-panneaux-solaires-application-au-cas-de-la-vigne-et-des-cultures-maraicheres-y-elamri.

European Food Safety Authority (2014). Outcome of the consultation with Member States and EFSA on the basic substance application for Rheum officinale and the conclusions drawn by EFSA on the specific points raised. *EFSA Support. Publ.*, 11 [Online]. Available at: https://doi.org/10.2903/sp.efsa.2014.EN-617.

Evans, W.C. (1976). Bracken thiaminase-mediated neurotoxic syndromes. *Bot. J. Linn. Soc.*, 73, 113–131 [Online]. Available at: https://doi.org/10.1111/j.1095-8339.1976.tb02017.x.

Évreux, C. (1989). La grande consoude, *Symphytum officinale* L., Borraginacées. Thesis, Joseph Fourier University, Grenoble [Online]. Available at: https://dumas.ccsd.cnrs.fr/dumas-01675622.

Fardeau, J.-C. and Jonis, M. (2004). AlterAgri, Phytostimulants et éliciteurs pour végétaux [Online]. Available at: https://ecophytopic.fr/sites/default/files/AA65_21-26_cle477246.pdf.

Foucault, Y., Lévêque, T., Xiong, T., Schreck, E., Austruy, A., Shahid, M., Dumat, C. (2013). Green manure plants for remediation of soils polluted by metals and metalloids: Ecotoxicity and human bioavailability assessment. *Chemosphere*, 93, 1430–1435 [Online]. Available at: https://doi.org/10.1016/j.chemosphere.2013.07.040.

Fu, J., Li, C.-H., Zhao, J.-R., Ma, L., Liu, T.-X. (2009). Shade-tolerance indices of maize: Selection and evaluation. *Yinggyong Shengtai Xuebao*, 20, 2705–2709.

Gabaston, J. (2018). Stilbènes de la vigne et d'essences forestières (pin, épicéa) : étude phytochimique et recherche d'activités anti-oomycète et insecticide. Thesis, University of Bordeaux, Bordeaux.

Gamm Vert (2018). Le purin d'ortie est-il dangereux ? [Online]. Available at: https://www.gammvert.fr/conseils/conseils-de-jardinage/le-purin-d-ortie-est-il-en-danger.

Ghedira, K. and Goetz, P. (2012). *Symphytum officinale* L. (Boraginaceae). *Phytothérapie*, 10, 382–386 [Online]. Available at: https://doi.org/10.1007/s10298-012-0745-7.

GIEC-IPCC (2014). Technology-specific cost and performance parameters. Appendix III, 5th GIEC-IPCC report.

Gindro, K., Godard, S., De Groote, I., Viret, O., Forrer, H.-R., Dornart, B. (2007). Peut-on stimuler les mécanismes de défense de la vigne ? Une nouvelle méthode pour évaluer le potentiel des éliciteurs. *Revue suisse Vitic. Arboric. Hortic.*, 39(6), 377–383.

Gonçalves, R.S., Silva, E.L., Hioka, N., Nakamura, C.V., Bruschi, M.L., Caetano, W. (2018). An optimized protocol for anthraquinones isolation from *Rhamnus frangula* L. *Nat. Prod. Res.*, 32, 366–369 [Online]. Available at: https://doi.org/10.1080/14786419.2017.1356836.

Greenpeace France (n.d). Quel est l'impact environnemental des panneaux solaires [Online]. Available at: https://www.greenpeace.fr/impact-environnemental-solaire/.

Griffiths-Lee, J., Nicholls, E., Goulson, D. (2020). Companion planting to attract pollinators increases the yield and quality of strawberry fruit in gardens and allotments. *Ecol. Entomol.*, 4(5), 1025–1034.

GuidEnR (n.d.). Photovoltaique, l'information voltaïque [Online]. Available at: http://www.photovoltaique.guidenr.fr/cours-photovoltaique-autonome-1/influence-localisation-gisement-solaire.php.

Hamdaoui, S. (2019). Florensac se lance dans l'agrivoltaïsme ou comment combiner agriculture et photovoltaïque. *France Bleu* [Online]. Available at: https://www.francebleu.fr/infos/agriculture-peche/florensac-se-lance-dans-l-agrivoltaisme-ou-comment-combiner-agriculture-et-photovoltaique-1563548297.

Hamelin, C., Gagnon, D., Truax, B. (2015). Aboveground biomass of glossy buckthorn is similar in open and understory environments but architectural strategy differs. *Forests*, 6, 1083–1093 [Online]. Available at: https://doi.org/10.3390/f6041083.

Hassanpour Adeh, E., Selker, J.S., Higgins, C.W. (2018). Remarkable agrivoltaic influence on soil moisture, micrometeorology and water-use efficiency. *PLOS One*, 13, e0203256 [Online]. Available at: https://doi.org/10.1371/journal.pone.0203256.

He, J., Wang, L., Guo, H., Zhao, H., Sun, J. (2019). The review of chemistry, pharmacology and processing method of rhubarb (*Rheum* species). *J. Food Bioact.*, 8 [Online]. Available at: https://doi.org/10.31665/JFB.2019.8205.

Hebrard, O. (2020). Face aux sécheresses, changer de modèle agricole. *Le Monde* [Online]. Available at: https://www.lemonde.fr/idees/article/2020/08/29/face-aux-secheresses-changer-de-modele-agricole_6050263_3232.html.

Hertwich, E.G., Gibon, T., Bouman, E.A., Arvesen, A., Suh, S., Heath, G.A., Bergesen, J.D., Ramirez, A., Vega, M.I., Shi, L. (2015). Integrated lifecycle assessment of electricity-supply scenarios confirms global environmental benefit of low-carbon technologies. *Proc. Natl. Acad. Sci.*, 112, 6277–6282.

Hirzel, J., Moya-Elizondo, E., Hernández, M., Guzmán, P., González, D. (2020). Effect of shade cloth on the evolution of nutrient concentrations in apple tree leaves. *Sci. Hortic.*, 266 [Online]. Available at: https://doi.org/10.1016/j.scienta.2020.109288.

Hsu, S.-C. and Chung, J.-G. (2012). Anticancer potential of emodin. *BioMedicine*, 2, 108–116 [Online]. Available at: https://doi.org/10.1016/j.biomed.2012.03.003.

Huang, Q., Lu, G., Shen, H.-M., Chung, M.C.M., Ong, C.N. (2007). Anti-cancer properties of anthraquinones from rhubarb. *Med. Res. Rev.*, 27, 609–630 [Online]. Available at: https://doi.org/10.1002/med.20094.

Hunt, A.J., Matharu, A.S., King, A.H., Clark, J.H. (2015). The importance of elemental sustainability and critical element recovery. *Green Chem.*, 17, 1949–1950 [Online]. Available at: https://doi.org/10.1039/C5GC90019K.

Inra (2017). L'Inra parle de l'agrivoltaïque. *Éco Solution Énergie* [Online]. Available at: https://www.ecosolutionenergie.com/actualites/inra-parle-de-l-agrivoltaique.

Institut technique de l'horticulture (2017). Liste de plantes attractives pour les abeilles – Plantes nectarifères et pollinifères à semer et à planter [Online]. Available at: https://www.valhor.fr/fileadmin/A-Valhor/Valhor_PDF/Liste_Plantes_attractives-abeilles.pdf.

Irigoyen, W. (2017). Chili : la Mecque de l'énergie solaire. *Arte* [Online]. Available at: https://www.arte.tv/fr/videos/079194-000-A/chili-la-mecque-de-l-energie-solaire/.

ITAB (2019). Préparation naturelle peu préoccupante [Online]. Available at: http://www.itab.asso.fr/activites/pnpp.php.

ITAB, GRAB, OPABA (2013). Journées techniques nationales fruits, légumes et viticulture biologiques [Online]. Available at: http://itab.asso.fr/downloads/actes3/actes-jtcolmar-web.pdf.

Jancovici, J.-M. and Grandjean, A. (2006). *Le plein s'il vous plaît : la solution au problème de l'énergie.* Le Seuil, Paris.

Jeandet, P., Sobarzo-Sánchez, E., Silva, A.S., Clément, C., Nabavi, S.F., Battino, M., Rasekhian, M., Belwal, T., Habtemariam, S., Koffas, M., Nabavi, S.M. (2020). Whole-cell biocatalytic, enzymatic and green chemistry methods for the production of resveratrol and its derivatives. *Biotechnol. Adv.*, 39, 107461 [Online]. Available at: https://doi.org/10.1016/j.biotechadv.2019.107461.

Kalisz, S., Oszmiański, J., Kolniak-Ostek, J., Grobelna, A., Kieliszek, M., Cendrowski, A. (2020). Effect of a variety of polyphenols compounds and antioxidant properties of rhubarb (*Rheum rhabarbarum*). *LWT*, 118, 108775 [Online]. Available at: https://doi.org/10.1016/j.lwt.2019.108775.

Kalyanpur, A., Mercadier, M.-E., Blanc, P. (2013). Gisement solaire en France caractérisation de la ressource énergétique, profil de répartition et volatilité. *Environnement & Technique*, 331, 54–59.

Kisielius, V., Lindqvist, D.N., Thygesen, M.B., Rodamer, M., Hansen, H.C.B., Rasmussen, L.H. (2020). Fast LC-MS quantification of ptesculentoside, caudatoside, ptaquiloside and corresponding pterosins in bracken ferns. *J. Chromatogr.*, B 1138, 121966 [Online]. Available at: https://doi.org/10.1016/j.jchromb.2019.121966.

Kosikowska, U., Smolarz, H., Malm, A. (2010). Antimicrobial activity and total content of polyphenols of *Rheum* L. species growing in Poland. *Open Life Sci.*, 5 [Online]. Available at: https://doi.org/10.2478/s11535-010-0067-4.

La France agricole (n.d.) Agrivoltaïsme : que valent les moutons sous les panneaux photovoltaïques [Online]. Available at: https://www.lafranceagricole.fr/actualites/elevage/agrivoltaisme-que-valent-les-moutons-sous-les-panneaux-photovoltaiques-1,13,271215027.html.

L'Écho du Solaire (2017). L'agriphotovoltaïque, c'est fantastique [Online]. Available at: https://www.lechodusolaire.fr/lagriphotovoltaique-cest-fantastique/.

L'Écho du Solaire (2018a). Le photovoltaïque bénéficie à l'élevage des esturgeons [Online]. Available at: https://www.lechodusolaire.fr/photovoltaique-beneficie-a-lelevage-desturgeons-aquitaine/.

L'Écho du Solaire (2018b). Photovoltaïque et agriculture font bon ménage [Online]. Available at: https://www.lechodusolaire.fr/photovoltaique-et-agriculture-font-bon-menage/.

L'Écho du Solaire (2018c). Serres solaires : une inauguration avec retour d'expérience sur un site de 18 m² en Isère [Online]. Available at: https://www.lechodusolaire.fr/serres-solaires-une-inauguration-avec-retour-dexperience-sur-un-site-de-18-mw-en-isere/.

L'Info durable (2019). En Provence, une carrière désaffectée devient la première centrale solaire flottante de France [Online]. Available at: https://www.linfodurable.fr/environnement/en-provence-une-carriere-desaffectee-devient-la-premiere-centrale-solaire-flottante.

Lapouge-Déjean, B., David, F., Lapouge, S. (2018). *Je cultive en lasagnes partout et toute l'année*. Terre Vivante, Mens.

Le contrôle climatique des cultures (2020). Ombrea [Online]. Available at: https://www.ombrea.fr/le-systeme.

Lee, M.S. and Sohn, C.B. (2008). Anti-diabetic properties of Chrysophanol and its glucoside from rhubarb rhizome. *Biol. Pharm. Bull.*, 31, 2154–2157 [Online]. Available at: https://doi.org/10.1248/bpb.31.2154.

Legifrance (2006). Loi n° 2006-11 d'orientation agricole.

Legifrance (2008). Article D4211-11. Code de la santé publique.

Legifrance (2015). Loi n° 2015-992 relative à la transition énergétique pour la croissance verte.

Li, C., Wang, H., Miao, H., Ye, B. (2017). The economic and social performance of integrated photovoltaic and agricultural greenhouses systems: Case study in China. *Appl. Energy*, 190, 204–212 [Online]. Available at: https://doi.org/10.1016/j.apenergy.2016.12.121.

Ligne des 400 (n.d.). Akuo Energy [Online]. Available at: https://www.akuoenergy.com/fr/ligne-des-400.

Majak, W. and Cheng, K.-J. (1987). Hydrolysis of the cyanogenic glycosides amygdalin, prunasin and linamarin by ruminal microorganisms. *Can. J. Anim. Sci.*, 67, 1133–1137 [Online]. Available at: https://doi.org/10.4141/cjas87-120.

Marrou, H., Guilioni, L., Dufour, L., Dupraz, C., Wery, J. (2013a). Microclimate under agrivoltaic systems: Is crop growth rate affected in the partial shade of solar panels? *Agric. For. Meteorol.*, 177, 117–132 [Online]. Available at: https://doi.org/10.1016/j.agrformet.2013.04.012.

Marrou, H., Wery, J., Dufour, L., Dupraz, C. (2013b). Productivity and radiation use efficiency of lettuces grown in the partial shade of photovoltaic panels. *Eur. J. Agron.*, 44, 54–66 [Online]. Available at: https://doi.org/10.1016/j.eja.2012.08.003.

Masson, P. (2018). *Guide pratique pour l'agriculture biodynamique*. BioDynamie Services, Chateau.

Metayer, N. (2020). LIFE AgriAdapt : adaptation durable des systèmes agricoles de l'Union européenne au changement climatique [Online]. Available at: https://agriadapt.eu/?lang=fr.

Ministère de l'Écologie, de l'Énergie, du Développement durable et de la Mer (2011). Évaluation du potentiel de production d'électricité d'origine solaire en Aquitaine.

Ministère de l'Écologie, du Développement durable, des Transports et du Logement (2011). Installations photovoltaïques au sol, Guide de l'étude d'impact [Online]. Available at: https://www.ecologie.gouv.fr/sites/default/files/Guide_EI_Installations-photovolt-au-sol_DEF_19-04-11.pdf.

Ministère de la Transition écologique et solidaire (2017). Des projets exemplaires de financement participatif dans les territoires [Online]. Available at: https://www.ecologique-solidaire.gouv.fr/des-projets-exemplaires-financement-participatif-dans-territoires.

Ministère de la Transition écologique et solidaire (2020a). Solaire [Online]. Available at: https://www.ecologie.gouv.fr/solaire.

Ministère de la Transition écologique et solidaire (2020b). Guide 2020 : L'instruction des demandes d'autorisations d'urbanisme pour les centrales solaires au sol.

Moerman, R., Roger, N., De Jonghe, R., Michez, D., Vanderplanck, M. (2016). Interspecific variation in bumblebee performance on pollen diet: New insights for mitigation strategies. *PLOS One*, 11, e0168462 [Online]. Available at: https://doi.org/10.1371/journal.pone.0168462.

Moine, G. (2016). *L'électrification solaire photovoltaïque*. ObservER, Paris.

Nieder, C., Wiedenmann, N., Andratschke, N., Molls, M. (2006). Current status of angiogenesis inhibitors combined with radiation therapy. *Cancer Treat. Rev.*, 32, 348–364 [Online]. Available at: https://doi.org/10.1016/j.ctrv.2006.03.006.

Now, D.B. (2019). Organic poultry farm in Hartly opts for solar. *Del. Bus. Now* [Online]. Available at: https://delawarebusinessnow.com/2019/01/organic-poultry-farm-in-hartly-opts-for-solar/.

Observatoire des énergies renouvelables (2019). Suivi du marché 2018 des installations solaires photovoltaïques individuelles. Study, Observ'ER.

O'Driscoll, C., Ramwell, C., Harhen, B., Morrison, L., Clauson-Kaas, F., Hansen, H., Campbell, G., Sheahan, J., Misstear, B., Xiao, L. (2016). Ptaquiloside in Irish bracken ferns and receiving waters, with implications for land managers. *Molecules*, 21, 543 [Online]. Available at: https://doi.org/10.3390/molecules21050543.

Peace, L.G. and Christopher, N.D. (2019). Evaluating the roles of biochar in climate change mitigations and soil amendments – A review. *Glob. Sci. J.*, 7, 27.

Pelt, J.M. (2011). *L'évolution vue par un botaniste*. Fayard, Paris.

Piotrowska, H., Kucinska, M., Murias, M. (2012). Biological activity of piceatannol: Leaving the shadow of resveratrol. *Mutat. Res. Mutat. Res.*, 750, 60–82 [Online]. Available at: https://doi.org/10.1016/j.mrrev.2011.11.001.

Pitron, G. (2018). *La guerre des métaux rares : la face cachée de la transition énergétique et numérique*. Les Liens qui libèrent, Paris.

Polese, J.-M. (2006). *La culture des plantes aromatiques*. Artemis, Chamalières.

Potter, D.M. and Baird, M.S. (2000). Carcinogenic effects of ptaquiloside in bracken fern and related compounds. *Br. J. Cancer*, 83, 914–920 [Online]. Available at: https://doi.org/10.1054/bjoc.2000.1368.

Pouliquen, F. (2020). L'agrivoltaïsme met les panneaux solaires au service des cultures. *20 minutes* [Online]. Available at: https://www.20minutes.fr/planete/2812707-20200710-agrivoltaisme-quand-panneaux-solaires-mettent-service-cultures.

Rasmussen, L.H. (2003). Ptaquiloside – An environmental hazard. Thesis, Royal Veterinary and Agricultural University, Thorvaldsensvej.

Rasmussen, L.H., Schmidt, B., Sheffield, E. (2013). Ptaquiloside in bracken spores from Britain. *Chemosphere*, 90, 2539–2541 [Online]. Available at: https://doi.org/10.1016/j.chemosphere.2012.10.092.

Rasmussen, L.H., Donnelly, E., Strobel, B.W., Holm, P.E., Hansen, H.C.B. (2015). Land management of bracken needs to account for bracken carcinogens – A case study from Britain. *J. Environ. Manage.*, 151, 258–266 [Online]. Available at: https://doi.org/10.1016/j.jenvman.2014.12.052.

Reinartz, J.A. and Kline, J. (1988). Glossy buckthorn (*Rhamnus frangula*), a threat to the vegetation of the Cedarburg Bog. *Field Station Bulletin*, 21(2), 20–35.

Roggemans, J. and Boxus, P. (1988). Rhubarb (*Rheum rhaponticum* L.). In *Crops II, Biotechnology in Agriculture and Forestry*, Bajaj, Y.P.S. (ed.). Springer, Berlin [Online]. Available at: https://doi.org/10.1007/978-3-642-73520-2_16.

Sekiyama, T. and Nagashima, A. (2019). Solar sharing for both food and clean energy production: Performance of agrivoltaic systems for corn. A typical shade-intolerant crop. *Environments*, 6, 65 [Online]. Available at: https://doi.org/10.3390/environments6060065.

Sentenac, G. and Thiery, D. (2008). *Les méthodes de lutte biologique ou biotechnique contre les insectes nuisibles à la vigne*. Mondiativi, Bordeaux.

Shaikh, S. and Patil, M.A. (2013). Production and utilization strategies of organic fertilizers for organic farming: An eco-friendly approach. *International Journal of Life Sciences and Pharma Research*, 3(2), 1–5.

Staiger, C. (2012). Comfrey: A clinical overview. *Phytother. Res.*, 26, 1441–1448 [Online]. Available at: https://doi.org/10.1002/ptr.4612.

Sun, H., Luo, G., Chen, D., Xiang, Z. (2016). A comprehensive and system review for the pharmacological mechanism of action of Rhein, an active anthraquinone ingredient. *Front. Pharmacol.*, 7 [Online]. Available at: https://doi.org/10.3389/fphar.2016.00247.

Tecsol blogs (2020). Agrivoltaïsme : BayWa r.e. met en service un projet de 2,7 MWc aux Pays-Bas avec des modules semi-transparents innovants. *Tescol Blogs* [Online]. Available at: https://tecsol.blogs.com/mon_weblog/2020/08/agrivolta%C3%AFsme-baywa-re-met-en-service-un-projet-de-27-mwc-aux-pays-bas-avec-des-modules-semi-transpa.html.

Tureckova, J. (2001). Rhubarb – A little known perennial vegetable. *Vyziva Potraviny Czech Repub.*, 56, 8–9.

UICN France, FBCN, AFB, MNHN (2018). Flore vasculaire de France métropolitaine. Liste Rouge Espèces Menacées en France [Online]. Available at: https://uicn.fr/wp-content/uploads/2019/01/liste-rouge-de-la-flore-vasculaire-de-france-metropolitaine.pdf.

Valle, B., Simonneau, T., Boulord, R., Sourd, F., Frisson, T., Ryckewaert, M., Hamard, P., Brichet, N., Dauzat, M., Christophe, A. (2017). PYM: A new, affordable, image-based method using a raspberry Pi to phenotype plant leaf area in a wide diversity of environments. *Plant Methods*, 13, 98 [Online]. Available at: https://doi.org/10.1186/s13007-017-0248-5.

Van Daele, C. (2011). *Plantes d'intérêt apicole*. Centre technique horticole de Gembloux, 5.

Van Leeuwen, C. and Darriet, P. (2016). The impact of climate change on viticulture and wine quality. *J. Wine Econ.*, 11, 150–167 [Online]. Available at: https://doi.org/10.1017/jwe.2015.21.

Verhaeghe, L. and Blais, A. (n.d.) Lendosphere : crowdfunding et transition énergétique [Online]. Available at: https://www.lendosphere.com/les-projets/toiture-photovoltaique-de-caviar-de-neuvic-en-dord.

Vetter, J. (2009). A biological hazard of our age: Bracken fern [*Pteridium aquilinum* (L.) Kuhn] – A review. *Acta Vet. Hung.*, 57, 183–196 [Online]. Available at: https://doi.org/10.1556/AVet.57.2009.1.18.

Vitisphere (2020). Premiers résultats concluants sur l'ombrage photovoltaïque [Online]. Available at: http://www.vitisphere.com/breve-91285-Premiers-resultats-concluants-sur-lombrage-photovoltaique.html.

Wajant, H., Riedel, D., Benz, S., Mundry, K.-W. (1994). Immunocytological localization of hydroxynitrile lyases from *Sorghum bicolor* L. and *Linum usitatissimum* L. *Plant Sci.*, 103, 145–154 [Online]. Available at: https://doi.org/10.1016/0168-9452(94)90202-X.

Will, F. and Dietrich, H. (2013). Processing and chemical composition of rhubarb (*Rheum rhabarbarum*) juice. *Food Sci. Technol.*, 50, 673–678 [Online]. Available at: https://doi.org/10.1016/j.lwt.2012.07.029.

Xue, J. (2017). Photovoltaic agriculture – New opportunity for photovoltaic applications in China. *Renew. Sustain. Energy Rev.*, 73, 1–9 [Online]. Available at: https://doi.org/10.1016/j.rser.2017.01.098.

Yamada, K., Ojika, M., Kigoshi, H. (1998). Isolation, chemistry, and biochemistry of Ptaquiloside, a bracken carcinogen. *Angew. Chem. Int. Ed.*, 37, 1818 1826.

Zhang, R., Kang, K.A., Piao, M.J., Lee, K.H., Jang, H.S., Park, M.J., Kim, B.J., Kim, J.S., Kim, Y.S., Ryu, S.Y., Hyun, J.W. (2007). Rhapontigenin from *Rheum undulatum* protects against oxidative-stress-induced cell damage through antioxidant activity. *J. Toxicol. Environ. Health*, A70, 1155–1166 [Online]. Available at: https://doi.org/10.1080/15287390701252766.

Zhao, J. (2011). The extraction of high value chemicals from heather (*Calluna vulgaris*) and bracken (*Pteridium aquilinum*). Thesis, University of York, York [Online]. Available at: http://etheses.whiterose.ac.uk/2019/2/Final_Thesis_of_Jiewen_Zhao.pdf.

Zheng, Q., Wu, H., Guo, J., Nan, H., Chen S., Yang, J., Xu, X. (2013). Review of rhubarbs: Chemistry and pharmacology. *Chinese Herbal Medicines*, 5, 932 [Online]. Available at: https://doi.org/10.750 1/j.issn.1674-6384.2013.01.003.

Index

Other titles from

ISTE

in

Ecological Science

2021

FLEURENCE Joël
Microalgae: From Future Food to Cellular Factory

AMIARD Jean-Claude
Disarmament and Decommissioning in the Nuclear Domain (Radioactive Risk SET – Volume 4)

JOLY Fernand, BOURRIÉ Guilhem
Mankind and Deserts 1: Deserts, Aridity, Exploration and Conquests
Mankind and Deserts 2: Water and Salts
Mankind and Deserts 3: Wind in Deserts and Civilizations

MUTTIN Frédéric, THOMAS Hélène
Marine Environmental Quality

2020

BRUSLÉ Jacques, QUIGNARD Jean-Pierre
Fish Behavior 1: Eco-ethology
Fish Behavior 2: Ethophysiology

LE FLOCH Stéphane
Remote Detection and Maritime Pollution: Chemical Spill Studies

MIGON Christophe, NIVAL Paul, SCIANDRA Antoine
The Mediterranean Sea in the Era of Global Change 1: 30 Years of Multidisciplinary Study of the Ligurian Sea
The Mediterranean Sea in the Era of Global Change 2: 30 Years of Multidisciplinary Study of the Ligurian Sea

ROSSIGNOL Jean-Yves
Climatic Impact of Activities: Methodological Guide for Analysis and Action

2019

AMIARD Jean-Claude
Industrial and Medical Nuclear Accidents: Environmental, Ecological, Health and Socio-economic Consequences (Radioactive Risk SET – Volume 2)
Nuclear Accidents: Prevention and Management of an Accidental Crisis (Radioactive Risk SET – Volume 3)

BOULEAU Gabrielle
Politicization of Ecological Issues: From Environmental Forms to Environmental Motives

DAVID Valérie
Statistics in Environmental Sciences

GIRAULT Yves
UNESCO Global Geoparks: Tension Between Territorial Development and Heritage Enhancement

KARA Mohamed Hichem, QUIGNARD Jean-Pierre
Fishes in Lagoons and Estuaries in the Mediterranean 2: Sedentary Fish
Fishes in Lagoons and Estuaries in the Mediterranean 3A: Migratory Fish
Fishes in Lagoons and Estuaries in the Mediterranean 3B: Migratory Fish

OUVRARD Benjamin, STENGER Anne
Incentives and Environmental Policies: From Theory to Empirical Novelties

2018

AMIARD Jean-Claude
Military Nuclear Accidents: Environmental, Ecological, Health and Socio-economic Consequences
(Radioactive Risk SET – Volume 1)

FLIPO Fabrice
The Coming Authoritarian Ecology

GUILLOUX Bleuenn
Marine Genetic Resources, R&D and the Law 1: Complex Objects of Use

KARA Mohamed Hichem, QUIGNARD Jean-Pierre
Fishes in Lagoons and Estuaries in the Mediterranean 1: Diversity, Bioecology and Exploitation

2016

BAGNÈRES Anne-Geneviève, HOSSAERT-MCKEY Martine
Chemical Ecology

2014

DE LARMINAT Philippe
Climate Change: Identification and Projections